RESEARCH AND PERSPECTIVES IN NEUROSCIENCES

Fondation Ipsen

Editor

Yves Christen, Fondation Ipsen, Paris (France).

Springer
Berlin
Heidelberg
New York
Barcelona
Hong Kong
London
Milan
Paris
Singapore
Tokyo

P. Patterson C. Kordon Y. Christen (Eds.)

Neuro-Immune Interactions in Neurologic and Psychiatric Disorders

With 43 Figures and 9 Tables

Springer

Prof. Dr. Paul Patterson
California Institute of Technology
Biology Division
Pasadena, CA 91125, USA

Dr. Ph. D. Claude Kordon
INSERM U159
Centre Paul Broca
2ter, Rue d'Alesia
75014 Paris, France

Ph. D. Yves Christen
Fondation IPSEN
24, Rue Erlanger
75781 Paris Cedex 16, France

ISSN 0945-6082
ISBN-13: 978-3-642-64065-0 e-ISBN-13: 978-3-642-59643-8
DOI: 10.1007/978-3-642-59643-8

Library of Congress Cataloging-in-Pulication Data
Neuro-immune, neurodegenerative and psychiatric disorders and neural injury/P. Patterson, C. Kordon, Y. Christen (eds.). p. cm. - (Research and perspectives in neurosciences, ISSN 0945-6082) Includes bibliographical references and index. ISBN 3-540-66013-5 (hardcover: alk. paper) 1. Nervous system - Degeneration - Immunological aspectcs. 2. Neuroimmunology. 3. Cytokines - Pathophysiology. 4. Alzheimer's disease - Immunological aspects. I. Patterson, Paul H. II. Kordon, Claude. III. Christen, Yves. IV. Title: Neuroimmune, neurodegenerative und pyschiatric disorders and neural injury. V. Series. [DNLM: 1. Neurodegenerative Diseases - immunology. 2. Cytokines - immunology. 3. Inflammation - immunology. WL 359 N4948 1999] RC365.N425 1999 616.8'0479 - dc21 DNLM/DLC for Library of Congress

Softcover reprint of the hardcover 1st edition 2000

Production: PRO EDIT GmbH, D-69126 Heidelberg
Cover design: Design & Production, D-69121 Heidelberg
Typesetting: K+V Fotosatz GmbH, D-64743 Beerfelden

SPIN 10716205 27/3136-5 4 3 2 1 0

Preface

The idea that the brain is an "immune-privileged site" has perhaps served to slow our realization that the intact brain can generate its own inflammatory reactions. These responses can be to peripheral infection, or they can arise from local, internal causes, for instance as a response to stress or to the severe changes in neuronal activity in seizure or the loss of oxygen in stroke. We are also becoming increasingly aware of the contribution of local inflammatory reactions to certain neurodegenerative diseases such as Alzheimer's disease (AD). In fact, evidence is accumulating that inflammatory processes contribute to the progression of AD, suggesting the possibility of using currently available or novel anti-inflammatory agents to interfere with this terrible disease. Correlations are also being made between inflammatory signs and mental illness, which is a new frontier of research.

This book presents the current state of knowledge in a variety of areas relevant to neuro-immune interactions, with particular attention to AD. These chapters summarize presentations made at the October 5, 1998 meeting of the Fondation IPSEN, in Paris. Inflammation has been linked to AD both from epidemiological work and from neuropathological studies of the brain. Correlations have been made between taking anti-inflammatory drugs and postponing the onset of AD, and a variety of inflammatory markers have been localized to the senile plaques, the latter being diagnostic for AD pathology. These and related topics are discussed by Joseph Rogers and Sue Griffin in this volume. The use of transgenic mice to study AD is described by Lennert Mucke and Paul Patterson, who present new findings on the interplay of cytokines and apolipoprotein E (ApoE), and the use of anti-inflammatory drugs in this animal model. Mucke also presented novel work on the use of inducible promoters in controlling amyloid precursor protein (APP) and ApoE in new mouse models. The involvement of microglia, the phagocytic inflammatory cells of the brain, in this inflammatory reaction is discussed by Barbara Cordell, who is investigating the role of cytokines in the interaction between neurons and glia in culture. Giulio Levi also reports on the control of microglial activation and the role of cAMP. The role of the inflammatory regulator NF-κB in the microglial response to APP is discussed by Steven Barger. He also considers the regulation of NF-κB in neurons. Evidence for increases in microglial and astrocytic activation as a function of aging is presented by Caleb Finch, who ties this to the increased frequency of AD with age.

The role of cytokines in glial activation in the injured brain is taken up by Georg Kreutzberg, and many of these issues are considered by Hans Wekerle and Patterson as well. Kreutzberg cites new findings on the use of imaging methods to visualize inflammation in the living human brain. Wekerle also discusses neuro-immune interaction in neurodegenerative disorders and mental illness, as does Michael Maes. This is another emerging area with potential therapeutic relevance. Maes discusses the role of serotonin in depression, and Joannes Reul reports on the role of the brain serotonin system in peripheral inflammation and in sickness behavior. Robert Dantzer reviewed the role of cytokines in sickness behavior, and presented evidence that primary afferent nerves mediate the interactions between cytokines inside and outside the blood brain barrier.

The poster session was also very lively, and a number of these presentations were selected for inclusion in this volume. Gary Wenk and colleagues present findings from an AD model that involves infusion of lipopolysaccharide into the basal forebrain. Beatrice Vernet-der Garabedian et al. provide data on the role of the meninges in producing the cytokine IL-1β during peripheral inflammation. Philippe Brachet and colleagues discuss how vitamin D can act as a hormone in the control of neuro-immune interactions. Sylvian Bauer et al. present evidence on the possible involvement of cytokines in the response of olfactory sensory neurons to injury.

Overall, one is struck by the amount of molecular detail that is becoming available from studies of neuro-immune interaction, and on so many different fronts, from injury and infection, to neurodegenerative disease and mental illness. Many of these molecular players are potential therapeutic targets for diseases and conditions for which there are few, if any, adequate treatments now available. Thus, this interdisciplinary area is providing insights both useful and startling concerning the normal and pathological brain.

Jacqueline Mervaillie and her staff at the Fondation IPSEN were marvelous hosts, as usual, displaying smoothly gracious and cultured organization before, during and following the meeting, both on land and on the Seine.

Autumn 1999

PAUL PATTERSON
CLAUDE KORDON
YVES CHRISTEN

Contents

List of Contributors

Aubert, A.
Neurobiologie intégrative INSERM U394, Rue Camille Saint-Saens, 33077 Bordeaux Cedex, France

Bales, K.
Eli Lilly & Company, Lilly Corporate Center, Indianapolis, IN 46285, USA

Barger, S.W.
Departments of Geriatrics, Anatomy, and Internal Medicine, University of Arkansas for Medical Sciences, Geriatric Research Education and Clinical Center, John L. McClellan Memorial Veterans' Affairs Medical Center, Little Rock, AR 72205, USA

Bauer, S.
ESA 5020 CNRS, Université Lyon-I. Villeurbanne, France

Benahmed, M.
U 407 INSERM, Hôpital Lyon-Sud, Pierre-Benite, France

Bluthe, R.M.
Neurobiologie intégrative INSERM U 394, Rue Camille Saint-Saens, 33077 Bordeaux Cedex, France

Brachet, P.
INSERM U 437, CHU-Hôtel-Dieu, 44093 Nantes, France

Buttini, M.
Gladstone Institute of Neurological Disease and Department of Neurology, University of California, San Francisco, CA 94141-9100, USA

Cordell, B.
Scios Incorporated, 820 West Maude Avenue, Sunnyvale, CA 94086, USA

Dantzer, R.
Neurobiologie Intégrative INSERM U 394, Rue Camille Saint-Saens, 33077 Bordeaux Cedex, France

Darcy, F.
INSERM U298, Centre Hospitalier Universitaire, 49033 Angers, France

Delhaye-Bouchaud, N.
Laboratoire de Neurobiologie du Développement,
Institut des Neurosciences (CNRS UMR 7624),
Université Pierre et Marie Curie, 9 quai Saint Bernard,
75005 Paris, France

Finch, C.E.
Andrus Gerontology Center and the Department of Biological Sciences,
USC, Los Angeles, CA 90089-191, USA

Flügel, A.
Department of Neuroimmunology, Max-Planck-Institute for Neurobiology,
82152 Martinsried, Germany

Garcion, E.
INSERM U298, Centre Hospitalier Universitaire, 49033 Angers, France

Griffin, W.S.T.
Department of Veterans' Affairs Medical Center and Department of Geriatrics,
University of Arkansas for Medical Sciences and Geriatric Research,
Education, and Clinical Center, John L. McClellan Memorial Veterans' Hospital,
4300 West Seventh Street, Little Rock, AR 72205, USA

Hauss-Wegrzyniak, B.
Arizona Research Laboratories, Division of Neural Systems,
Memory and Aging, University of Arizona, Tucson, AZ 85724, USA

Jourdan, F.
ESA 5020 CNRS, Université Lyon-I. Villeurbanne, France

Kelley, K.W.
Laboratory of Immunophysiology, Department of Animal Science,
University of Illinois at Urbana, USA

Konsman, J.P.
Neurobiologie Intégrative INSERM U394, Rue Camille Saint-Saens,
33077 Bordeaux Cedex, France

Kreutzberg, G.W.
Max-Planck-Institute of Neurobiology,
Department of Neuromorphology, 82152 Martinsried, Germany

Kubera, M.
Clinical Research Center for Mental Health (CRC-MH),
University Department of Psychiatry, AZ Stuivenberg,
267 Lange Beeldekensstraat, 2060 Antwerp, Belgium

LANZREIN, A.-S.
Andrus Gerontology Center and the Department of Biological Sciences,
USC, Los Angeles, CA 90089-191, USA

LAYE, S.
Neurobiologie Intégrative INSERM U394, Rue Camille Saint-Saens,
33077 Bordeaux Cedex, France

LEMAIGRE-DUBREUIL, Y.
Laboratoire de Neurobiologie du Développement,
Institut des Neurosciences (CNRS UMR 7624),
Université Pierre et Marie Curie, 9 quai Saint Bernard,
75005 Paris, France

LEVI, GIULIO
Laboratory of Organ and System Pathophysiology,
Istituto Superiore di Sanità, Viale Regina Elena 299, 00161 Rome, Italy

LINTHORST, A.C.E.
Max-Planck-Institute of Psychiatry, Section Neuropsychopharmacology,
Kraepelinstr. 2–10, 80804 Munich, Germany

MAES, M.
Department of Psychiatry and Neuropsychology,
University Hospital of Maastricht, Postbus 5800,
6202 AZ Maastricht, The Netherlands

MAHLEY, R.W.
Gladstone Institute of Neurological Disease, Departments of Medicine
and Pathology, and Cardiovascular Research Institute,
University of California, San Francisco, CA 94141-9100, USA

MAO, X.
Department of Anatomy, University of Arkansas for Medical Sciences,
Little Rock, AR 72205, USA

MARIANI, J.
Laboratoire de Neurobiologie du Développement,
Institut des Neurosciences (CNRS UMR 7624),
Université Pierre et Marie Curie, 9 quai Saint Bernard,
75005 Paris, France

MAUDUIT, C.
U407 INSERM, Hôpital Lyon-Sud, Pierre-Benite, France

MOERMAN, A.M.
Department of Geriatrics, University of Arkansas for Medical Sciences,
Little Rock, AR 72205, USA

Montero-Menei, C.N.
INSERM U298, Centre Hospitalier Universitaire, 49033 Angers, France

Morgan, T.E.
Andrus Gerontology Center and the Department of Biological Sciences, USC, Los Angeles, CA 90089-191, USA

Moyse, E.
ESA 5020 CNRS, Université Lyon-I. Villeurbanne, France

Mrak, R.E.
Department of Veterans' Affairs Medical Center
and Department of Pathology,
University of Arkansas for Medical Sciences, Little Rock, AR 72205, USA

Mucke, L.
Gladstone Institute of Neurological Disease and Department of Neurology, University of California, San Francisco, CA 94141-9100, USA

Murphy, G.
Stanford University School of Medicine, Department of Psychiatry, Stanford, CA 94305, USA

Nataf, S.
INSERM U298, Centre Hospitalier Universitaire, 49033 Angers, France

Naidu, A.
Scios Incorporated, 820 West Maude Avenue, Sunnyvale, CA 94086, USA

Naveilhan, P.
INSERM U298, Centre Hospitalier Universitaire, 49033 Angers, France

Neumann, H.
Department of Neuroimmunology,
Max-Planck-Institut for Neurobiology, 82152 Martinsried, Germany

Neveu, I.
INSERM U298, Centre Hospitalier Universitaire, 49033 Angers, France

Parnet, P.
Neurobiologie Intégrative INSERM U394, Rue Camille Saint-Saens, 33077 Bordeaux Cedex, France

Patterson, P.H.
Biology Division, California Institute of Technology, Pasadena, CA 91125, USA

Paul, S.
Eli Lilly & Company, Lilly Corporate Center, Indianapolis, IN 46285, USA

PITAS, R.E.
Gladstone Institute of Neurological Disease, Department of Pathology and Cardiovascular Research Institute, University of California, San Francisco, CA 94141-9100, USA

RABER, J.
Gladstone Institute of Neurological Disease and Department of Neurology, University of California, San Francisco, CA 94141-9100, USA

RANGANATHAN, A.
Department of Anatomy, University of Arkansas for Medical Sciences, Little Rock, AR 72205, USA

REUL, J.M.H.M.
Max-Planck-Institute of Psychiatry, Section Neuropsychopharmacology, Kraepelinstr. 2–10, 80804 Munich, Germany

ROGERS, J.
Roberts Center for Alzheimer's Research, Sun Health Research Institute, 10515 West Santa Fe Drive, P.O. Box 1278, Sun City, AZ 85372, USA

ROZOVSKY, I.
Andrus Gerontology Center and the Department of Biological Sciences, USC, Los Angeles, CA 90089-191, USA

SHEN, Y.
Roberts Center for Alzheimer's Research, Sun Health Research Institute, 10515 West Santa Fe Drive, P.O. Box 1278, Sun City, AZ 85372, USA

SHENG, J.G.
Department of Geriatrics, University of Arkansas for Medical Sciences, Little Rock, AR 72205, USA
and Department of Neurology, Rui-Jin-Hospital, Shanghai Second Medical University, Shanghai 200025, China

SINDJI, L.
INSERM U298, Centre Hospitalier Universitaire, 49033 Angers, France

STONE, D.
Andrus Gerontology Center and the Department of Biological Sciences, USC, Los Angeles, CA 90089-191, USA

VERNET-DER GARABEDIAN, B.
Laboratoire de Neurobiologie du Développement, Institut des Neurosciences (CNRS UMR 7624), Université Pierre et Marie Curie, 9 quai Saint Bernard, 75005 Paris, France

WEKERLE, H.
Department of Neuroimmunology,
Max-Planck-Institut for Neurobiology, 82152 Martinsried, Germany

WENK, G. L.
Division of Neural Systems, University of Arizona,
384 Life Sciences North Building, Tucson, AZ 85724, USA

WILLARD, L. B.
Arizona Research Laboratories, Division of Neural Systems,
Memory and Aging, University of Arizona, Tucson, AZ 85724, USA

WION, D.
INSERM U298, Centre Hospitalier Universitaire, 49033 Angers, France

WYSS-CORAY, T.
Gladstone Institute of Neurological Disease
and Department of Neurology, University of California,
San Francisco, CA 94141-9100, USA

XIE, Z.
Andrus Gerontology Center and the Department of Biological Sciences,
USC, Los Angeles, CA 90089-191, USA

YANG, L.
Stanford University School of Medicine, Department of Psychiatry,
Stanford, CA 94305, USA

Organization of Inflammatory Processes in Alzheimer's Disease

J. Rogers and Y. Shen

Summary

Basic research and clinical studies have established inflammation as one of several important pathogenic mechanisms in Alzheimer's disease (AD). The sources of inflammation appear to be aggregated amyloid β peptide deposits, neurofibrillary tangles, and neurodegeneration itself. Organization of the inflammatory mediators and processes that have been uncovered in AD is likely to follow patterns similar to those observed in the periphery.

Over the last decade a virtual textbook of inflammatory mediators has been observed in the Alzheimer's disease (AD) brain. They are almost universally increased in expression in comparison to brain samples from the normal elderly (ND), and they are most often concentrated in regions of the brain that exhibit selective AD vulnerability (reviewed in Rogers and Griffin 1997; Rogers and O'Barr 1996; Rogers et al. 1996).

In broad scope how these inflammatory molecules and processes are organized should not be a great mystery. We have nearly a century of information from peripheral immunology that links, for example, complement to cytokines and cytokines back to complement. By judicious review of the same textbooks that we used as undergraduate and graduate students, we should therefore be able to arrive at a reasonable approximation of the way inflammation works in the AD brain.

Sources of AD Brain Inflammation

Inflammation, whether peripheral or central, almost always constitutes a reaction to something; it is typically a secondary event and seldom an underlying etiology. In AD research, where great battles have been fought over which of multiple mechanisms is primary and where the assertion that a mechanism is not primary relegates it to an epiphenomenon, lack of appreciation for the pathogenic significance of secondary inflammatory reactions initially created substantial obstacles. Consider, then, the plight of a patient who has been hit in the head with a bat. The etiologic event is the blow to the head, which doubtless has killed a few brain cells. What the patient's neurologist is concerned about, however, is not the primary event but the

Patterson/Kordon/Christen (Eds.)
Neuro-Immune Interactions in Neurologic and Psychiatric Disorders

secondary inflammatory reaction that may very well destroy so much of the brain that the patient dies. Secondary or not, inflammation should never be considered an epiphenomenon. This is especially true in the brain, one of the organs in the body most sensitive to inflammation.

What might the inflammation of AD be a reaction to? Again, this should not be a great mystery. Inflammation typically arises as a reaction to either an entity that is not normally present in a tissue or to degeneration within the tissue or both. A further clue is that inflammation appears to begin at early, and perhaps even preclinical, stages of the disease and to continue throughout the disease course (Lue et al. 1996a). Aggregated amyloid β peptide (Aβ) deposits and neurofibrillary tangles are pathologic entities not normally found in the brain (at least not in the profusion characteristic of AD). They are present from the very earliest preclinical phases to the terminal phases of AD. As the underlying cause of AD dementia, neurodegeneration also occurs throughout much of the disease process. These elements, then – aggregated Aβ deposits, neurofibrillary tangles, and neurodegeneration – are the most likely sources for AD inflammation and, as will be seen, there is now much basic science data that confirm such an hypothesis.

Aggregated Aβ and Neurofibrillary Tangles Activate Complement

We initially showed that Aβ activates the classical complement pathway in an antibody-independent manner (Rogers et al. 1992). This mechanism has been widely replicated and extended to reveal several critical details. Binding occurs between Aβ and the first component of the classical cascade, C1q, at a 13–15 amino acid sequence on the C1q A chain collagen tail (Jiang et al. 1994). This sequence is, in fact, a site for binding of most, if not all, antibody-independent activators of the classical pathway (Jiang et al. 1992). It is characterized by a predominance of exposed cationic side chains. The binding site for C1q on Aβ occurs at the Aβ N-terminus, a region that features at least four anionic side chains. C1q/Aβ binding is predictably altered by manipulation of salt concentrations, indicative of charge-based coupling (Webster et al. 1997b).

C1q is a hexamer made up of six identical subunits. As such, it proffers six Aβ binding sites per molecule. At the other end of the molecule, the globular head region, six binding sites are available for the binding of antibodies. Notably, binding and activation of C1q are increased by several orders of magnitude when multiple antibody binding sites of the same C1q molecule are occupied. This phenomenon may also apply at the antibody-independent site. Our structural studies, for example, suggest that the distances between adjacent antibody-independent sites on C1q and adjacent Aβ moieties when Aβ is in its aggregated, cross-β-pleated configuration are very close (Webster at al. 1995). Perhaps not surprisingly, the more aggregated Aβ becomes, the better activator it is (Webster et al. 1997a).

In studies our laboratory is now just concluding (Rogers et al., in preparation), neurofibrillary tangles, which become exposed extracellularly as "tombstone" tangles, have also been found to activate complement and to do so even more potently than Aβ. The tangle constituent that is responsible appears to be tau. Moreover, consistent with C1q structure and its interactions with Aβ, aggregates of tau – especially aggregates that may conform to the native configuration of tau within tangles – are better activators than freshly solubilized, presumably more monomeric tau.

The antibody-independent activation of complement in AD brain is unusual but by no means unprecedented. Bacteria and several RNA tumor viruses, for example, can stimulate complement in an antibody-independent fashion. Evidence for autoreactive antibodies or the B cells that produce them remains equivocal at best. We have used a dozen pan-B cell, anti-IgG, and anti-IgM antibodies and found patchy staining, typically near large, severed blood vessels, which was equally evident in AD, nondemented elderly, and normal young adult patients. Similarly, we were unable to observe immunoreactive natural killer cells; neither was there any obvious difference in the numbers or ratios of $CD4^+$ or $CD8^+$ T lymphocytes in AD and control brains (Rogers and Styren, unpublished).

Complement component C1q is highly co-localized with two pathologic entities in the AD brain: aggregated Aβ deposits and neurofibrillary tangles (Rogers et al. 1992). This finding, plus the clear mechanisms of Aβ and tangle stimulation of complement that have been demonstrated in vitro, makes complement activation by aggregated Aβ deposits and neurofibrillary tangles a logical source and starting point for AD inflammation. It is of the utmost importance to note, however, that inflammatory mechanisms are among the most highly interactive in the body. For example, complement can stimulate the production of cytokines and cytokines can stimulate the production of complement. Under these circumstances, the activation of the classical pathway by Aβ and tangles as a starting point for AD inflammation can only be regarded as a pedagogic convenience. In fact, Aβ also activates the alternative complement pathway (Bradt et al. 1998), stimulates cytokine production (Rogers and Griffin 1997), and has a host of other pro-inflammatory effects that may feed back to induce complement production and activation. For this reason, there should never be any semblance of an Aβ/tau war in the field of AD inflammation, for no AD inflammatory mechanism can be said to be more primary or more important than any other.

Generation of Opsonins and Anaphylatoxins

The activation of C1q leads to a complex cascade of mostly autocatalytic, serine protease-mediated reactions. The simplified classical pathway sequence is C1q, C1r/C1s, C4, C2, C3, C5, C5b-9. Cleavage products at most of the steps are generated at the site of initial activation. For this reason, the co-localization of complement activation fragments in the same place, aggregated Aβ

deposits and neurofibrillary tangles, provides strong presumptive evidence that full complement activation has occurred there. Some complement cleavage products, the opsonins C3b and iC3b, can bind back to the activator to help target scavenger cell actions. Others, anaphylatoxins such as C3a and C5a, provide chemotactic and stimulatory signals that activate and further guide inflammatory cells to the vicinity. Microglia in the AD brain possess receptors for some of these molecules. Our studies of microglial cultures from rapid autopsies of AD brain (Lue et al. 1996b) show consistent anaphylatoxin and opsonin responses to aggregated Aβ deposits in the wells, with microglia migrating to the deposits, becoming activated (e.g., MHCII-expressive), and phagocytosing the Aβ (Lue et al., in preparation). Likewise, in the AD brain microglial activation and clustering at aggregated Aβ deposits are well-known phenomena (Styren et al. 1990; Luber-Narod and Rogers 1988; Rogers et al. 1988). Although other mechanisms (e.g., RAGE, the macrophage scavenger receptor) also underlie microglial chemotaxis to Aβ, the role of complement should not be overlooked.

Production of Cytokines and Reactive Oxygen Species (ROS)

AD microglia in our cultures not only become activated, migrate to Aβ, and phagocytose it, but they also increase their production of cytokines, chemokines, complement, and ROS. These latter factors include IL-1β, TNF-α, IL-6, MIP-1α, MCP-1, and C1q (Lue et al., in preparation). Enhanced production of C1q, of course, suggests the first of many potential vicious cycle interactions of inflammatory mediators with AD pathology. For example, C1q is not only activated by Aβ but also (and for similar structural reasons) is the most potent enhancer of Aβ aggregation known (Webster and Rogers 1996). In turn, the more aggregated Aβ becomes, the better activator of complement it is (Webster et al. 1997a).

Because Aβ activation of complement is a major source and nidus for microglial activation, migration, Aβ phagocytosis, and production of inflammatory mediators, it is not surprising that all these processes primarily colocalize with aggregated Aβ deposits in the AD brain. The same is likely to prove true for extracellular neurofibrillary tangles, although such work has not yet received sufficient emphasis.

Actions of Cytokines, Chemokines and ROS

The actions and the identities of inflammatory mediators generated at sites of Aβ deposition and neurofibrillary tangle formation would warrant several chapters on each alone. For our purposes here it may be sufficient simply to note that, in addition to their cytotoxic actions, most of the pro-inflammatory cytokines are known to induce each other (especially TNF-α, IL-1, and IL-6), complement, and ROS, prompting further inflammatory cycles.

It may also be worth pointing out that the role of ROS in AD, studies of which have primarily centered on neuronal generation of ROS and its subsequent actions (e.g., apoptosis), almost certainly extends beyond this limited horizon. In fact, outside AD research ROS mechanisms are more likely to be encountered in the context of inflammation, where they constitute an important weapon in the armament of cytotoxic inflammatory cells.

Formation and Actions of the Membrane Attack Complex (MAC)

The MAC is the terminal component of both the classical and alternative pathways. It is made up of fragments and components of C5, C6, C7, C8, and C9, hence the alternative designation C5b-9. The final C5b-9 complex is a ring-like structure that can incorporate multiple C9 molecules, increasing the size of the pore formed by the MAC on targeted membranes and increasing the probability of cell lysis. In the periphery and in oligodendrocytes under MAC attack, the cells defend themselves by characteristic blebbing and endocytosis of those regions of their membranes where the MAC is fixed.

Although there was some initial controversy as to whether the MAC is formed in AD brain, recent studies may have provided a consensus that it is (Webster et al. 1997c). Multiple MAC antibodies label the same elements in the AD brain: plaques and tangles. Although these same plaques and tangles are also immunoreactive for antibodies directed against earlier classical pathway components, the staining patterns are different from those of the MAC: antibodies to the early complement components (e.g., C4d) label Aβ, whereas MAC antibodies appear to label cellular processes within the plaque (Rogers et al. 1992). This finding is completely consistent with the binding characteristics of early complement components, which, as previously noted, can bind back to the activator, and the MAC, which requires a membrane target. MAC immunostaining is scant or absent in ND patients and, even in AD patients, is scant or absent in brain structures such as cerebellum that exhibit little AD pathology. Western blots using an anti-MAC antibody show immunoreactive bands, matched by a MAC standard, that are dramatically enhanced in AD compared to ND cortical homogenates. Molecular studies demonstrate that mRNAs for most of the MAC constituents are present in brain. Perhaps most important, ultrastructure of AD plaques with anti-MAC antibodies reveals axons with MAC affixed to their membranes and characteristic blebbing and endocytotic responses at precisely the sites of MAC fixation (Webster et al. 1997c). These micrographs not only confirm that the MAC is present in the AD brain but also demonstrate that the MAC can actively damage neuronal processes there, not simply remove them as the detritus of other, non-inflammatory toxic mechanisms.

Defense Mechanisms

In addition to blebbing and endocytosis of the MAC, cells under inflammatory attack in the periphery exert molecular defenses. The best studied of these in AD, C4 binding protein and apolipoprotein J (apo J, SGP40-40; Johnson et al. 1992) show significant increases in the AD compared to the ND cortex (reviewed in Rogers and Griffin 1997; Rogers and O'Barr 1996; Rogers et al. 1996). Again, the very existence of this upregulation is strong presumptive evidence that active inflammatory attack is ongoing since, to our knowledge, there is no known instance of complement defense upregulation in the periphery without a significant pro-inflammatory stimulus.

CD59, a complement defense factor that blocks MAC insertion, presents something of a paradox for MAC actions in the AD brain. Its unique and highly practical feature is that it blocks homologous but not heterologous MAC, thereby warding off self-inflicted bystander lysis but permitting lysis of foreign targets. In the face of active MAC neurotoxicity in the AD brain, an important issue is, therefore, why CD59 is not more effective in the disorder. Perhaps there is a CD59 deficiency in AD brain. Alternatively, it is important to recognize that AD inflammation is not toxic shock syndrome. The latter can be lethal in a matter of hours, whereas AD cumulates damage over as much as a decade. Indeed, given the sensitivity of the brain to inflammation, it is likely that AD inflammation is relatively weak. Bystander lysis, despite CD59 upregulation, is an established phenomenon in inflammation biology. Perhaps, then, over the course of years MAC and other inflammatory mechanisms of attack elude common defense mechanisms in a small but cumulative number of instances sufficient to render micro-localized but chronic damage to the brain.

Acknowledgments. Our research is supported by grants from the National Institute on Aging (AGO7367) and a Zenith Award from the Alzheimer's Association.

References

Bradt BM, Kolb WP, Cooper NR (1998) Complement-dependent pro-inflammatory properties of the Alzheimer's disease β peptide. J Exp Med 188:431–438.

Jiang H, Robey FA, Gewurz H (1992) Localization of sites through which C-reactive protein binds and activates complement to residues 14–26 and 76–92 of the human C1q A chain. J Exp Med 175:1373–1379.

Jiang H, Burdick D, Glabe CG, Cotman CW, Tenner AJ (1994) β-amyloid activates complement by binding to a specific region of the collagen-like domain of the C1q chain. J Immunol 152: 5050–5059.

Johnson SA, Lampert-Etchells M, Pasinetti GM, Rozovsky I, Finch CE (1992) Complement mRNA in the mammalian brain: responses to Alzheimer's disease and experimental brain lesioning. Neurobiol Aging 13:641–648.

Luber-Narod J, Rogers J (1988) Immune system associated antigens expressed by cells of the human central nervous system. Neurosci Lett 94:17–22.

Lue, L-F, Brachova L, Civin WH, Rogers J (1996a) Inflammation, Aβ deposition, and neurofibrillary tangle formation as correlates of Alzheimer's disease neurodegeneration. J Neuropathol Exp Neurol 55:1083–1088.

Lue L-F, Brachova L, Walker DG, Rogers J (1996b) Characterization of glial cultures from rapid autopsies of Alzheimer's and control patients. Neurobiol Aging 17:421–429.

Rogers J, Cooper NR, Schultz J, McGeer PL, Webster S, Styren SD, Civin WH, Brachova L, Bradt B, Ward P, Lieberburg I (1992) Complement activation by β-amyloid in Alzheimer's disease. Proc Nat Acad Sci 89: 10016–10020

Rogers J, Griffin WST (1997) Inflammatory mechanisms of Alzheimer's disease: basic: research, clinical studies, and future directions. In: Wood P (ed) Neuroinflammation: mechanisms and management. Humana Press, Totawa, New Jersey, pp 177 – 193.

Rogers J, O'Barr S (1996) Inflammatory mediators in Alzheimer's disease. In: Tanzi R, Wasco W (eds) Molecular approaches to Alzheimer's disease. Humana Press, Totawa, New Jersey, pp 177–197.

Rogers J, Luber-Narod J, Styren SD, Civin WH (1988) Expression of immune system associated antigens by cells of the human central nervous system: relationship to the pathology of Alzheimer's disease. Neurobiol Aging 9:339–349.

Rogers J, Webster S, Lue L-F, Brachova L, Civin WH, Emmerling M, Shivers B, Walker D, McGeer PL (1996) Inflammation and Alzheimer's disease pathogenesis. Neurobiol Aging 17:681–686.

Styren SD, Civin WH, Rogers J (1990) Molecular, cellular, and pathologic characterization of HLA-DR immunoreactivity in normal elderly and Alzheimer's disease brain. Exp Neurol 110:93–104.

Webster S, Glabe C, Rogers J (1995) Multivalent binding of complement protein C1q to the amyloid β-peptide (Aβ) promotes the nucleation phase of Aβ aggregation. Biochem Biophys Res Commun 217:869–875.

Webster S, Bradt B, Rogers J, Cooper NR (1997a) Aggregation state-dependent activation of the classical complement pathway by the amyloid β peptide (Aβ). J Neurochem 69:388–398.

Webster S, Bonnell B, Rogers J (1997b) Charge based binding of complement component C1q to the Alzheimer amyloid β peptide. Am J Pathol 150:1531–1536.

Webster S, Lue L-F, Brachova L, Tenner A, McGeer PL, Walker D, Bradt B, Cooper NR, Rogers J (1997c) Molecular and cellular characterization of the membrane attack complex, C5b-9, in Alzheimer's disease. Neurobiol Aging 18:415–421.

Webster S, Rogers J (1996) Relative efficacies of amyloid β peptide (Aβ) binding proteins in Aβ aggregation. J Neurosci Res 46:58–66.

Glial Cytokines in Neurodegenerative Conditions*

R.E. Mrak, J.G. Sheng, and W.S.T. Griffin

Summary

The role of neuronal injury and glial inflammatory responses in Alzheimer's disease has been highlighted by recent epidemiological work establishing head trauma as an important risk factor and by the use of anti-inflammatory agents as an important ameliorating factor in this disease. Our studies have advanced the hypothesis that chronic activation of glial inflammatory processes is engendered by genetic or environmental insults to neurons, via neuronal signaling of microglia to become activated and to excessively elaborate the pro-inflammatory cytokine interleukin-1. This step sets in motion an interleukin-1 driven cytokine cycle of cellular and molecular events that is at once attractive in the short term, because of potential neuronotrophic consequences, and harmful in the long term, because of chronic neurodegenerative consequences. In this cycle, interleukin-1 1) promotes excessive synthesis and processing of the β-amyloid precursor protein and other β-amyloid plaque-associated proteins and 2) activates astrocytes with promotion of astrocytic synthesis and release of a number of inflammatory and neuroactive molecules. One of these interleukin-1-regulated astrocyte-derived molecules, S100β, is neuronotrophic. However, S100β may stress neurons, perhaps old ones in particular, as it promotes neurite growth and, like interleukin-1, induces excessive synthesis of neuronal β-amyloid precursor protein. In addition, S100β induces increases in intraneuronal free calcium concentrations that are neuronotoxic. The resulting neuronal injury can further activate microglia, with further overexpression of interleukin-1, and thereby produce feedback amplification and self-propagation of neurodegenerative processes. Unremitting self-propagation thus provides an explanation for progression of the neurodegenerative changes culminating in Alzheimer's disease.

* This work was supported in part by National Institutes of Health grants AG12411, AG10208, and NS27414.

Patterson/Kordon/Christen (Eds.)
Neuro-Immune Interactions in Neurologic and Psychiatric Disorders

Introduction

An important role for glia and glia-derived cytokines in the neurodegeneration of Alzheimer's disease has become apparent over the past decade. Microglia, astrocytes, and their cytokines *(e.g.,* interleukin-1 [IL-1] and S100β) are now recognized as active participants in a self-propagating cycle of neurodegeneration that underlies the progression of Alzheimer's disease (Griffin et al. 1998a) This cytokine-driven cycle leads to further neuronal injury, amyloid deposition, and transformation of amyloid deposits into neuritic β-amyloid plaques. The spread of disease from region to region of cerebral cortex can be ascribed to remote (neuronal target area) glial responses to neuronal or neuritic injury and degeneration, resulting in the spread of Alzheimer-type neuropathological changes to these remote areas. The role of chronic glial inflammatory-like processes, such as we describe here, is reflected in recent epidemiological observations suggesting beneficial effects of chronic anti-inflammatory therapy on the incidence and progression of Alzheimer's disease (Andersen et al. 1995; Breitner et al. 1994, 1995; Rich et al. 1995).

Glial Cytokines and Inflammation-like Reactions in Brain

In the periphery, IL-1 is produced by macrophages in response to acute insults, and drives a number of systemic inflammatory processes. Under conditions of incomplete resolution or continuing insult, this overexpression of IL-1 can become chronic, with detrimental consequences (Dinarello and Wolff 1993). In the central nervous system, IL-1 is also produced in response to injury by resident microglia (Giulian and Lachman 1985; Hetier et al. 1988; Griffin et al. 1994), and IL-1-orchestrated molecular and cellular cascades (Giulian et al. 1988; Sebire et al. 1993; Sheng et al. 1996a) in the central nervous system mirror IL-1-driven inflammation and repair processes in the periphery (Griffin et al. 1989; Rothwell 1991). As is the case in the periphery, chronic overexpression of IL-1 resulting from continuing or repeated insult is potentially damaging.

IL-1 manifests both autocrine and paracrine effects on cells of the central nervous system (Rothwell 1991). Autocrine effects of potential import for neurodegenerative processes include proliferation of, and further overexpression of IL-1 by, microglia (Ganter et al. 1992; Lee et al. 1993; Sebire et al. 1993). Paracrine effects of potential import include neuronal overexpression (Forloni et al. 1992; Goldgaber et al. 1989) and processing (Buxbaum et al. 1992) of the β-amyloid precursor protein (β-APP), and astrocytic activation (Giulian et al. 1988) and overexpression of plaque-associated proteins such as apolipoprotein E (ApoE; Das and Potter 1995) and S100β (Sheng et al. 1996a, see below). In addition, IL-1 has been shown to be directly neurotoxic at high levels in vitro (Brenneman et al. 1992; Strijbos and Rothwell 1995).

A particularly important paracrine effect of IL-1 is astrocytic overexpression of S100β (Sheng et al. 1996a). Overexpression of S100β has several important neurotrophic and neuritotrophic effects, including increased neurite outgrowth (Kligman and Marshak 1985) and elevated intraneuronal free calcium concentrations (Barger and van Eldik 1992). S100β also has autocrine effects on astrocytes, including astrocyte proliferation (Selinfreund et al. 1991), changes in astrocyte morphology (Selinfreund et al. 1990), and elevated intracellular free calcium concentrations (Barger and van Eldik 1992).

Glial Cytokines and Neurodegeneration in Alzheimer's Disease

Alzheimer's disease is characterized by striking and well-documented overexpression of IL-1 within affected cerebral cortical regions (Cacabelos et al. 1994; Griffin et al. 1989, 1995a; Sheng et al. 1995). There are both increased numbers of activated, IL-l-immunoreactive (IL-1^+) microglia and elevated tissue levels of IL-1. Similar intense, chronic microglial activation with overexpression of IL-1 is observed in Down's syndrome (Griffin et al. 1989), in which Alzheimer's disease is virtually inevitable by middle age (Wisniewski et al. 1985). In these patients, glial activation and glial cytokine overexpression is seen not only in middle-aged patients who show florid Alzheimer-type pathological changes, but also in young and even fetal Down's patients (Griffin et al. 1989, 1998b). These results suggest that activation of glia and elaboration of glial cytokines occurs years or even decades prior to the appearance of Alzheimer-type neuropathological changes and Alzheimer-related clinical symptoms.

In Alzheimer's disease itself, activated microglia overexpressing IL-1 are found in association both with neurons bearing neurofibrillary tangles (Sheng et al. 1997) and with extracellular amyloid plaques (Griffin et al. 1995a). Temporal patterns of glial association with amyloid plaques can be inferred from analysis of the distribution of activated microglia, overexpressing IL-1, among different plaque types representing stages in a hypothesized sequence (Rozemuller et al. 1989) of plaque evolution (Griffin et al. 1995a). Early, diffuse amyloid deposits without neuritic components contain such activated microglia in small numbers, suggesting an early role for IL-1 in the transformation of these amyloid deposits into neuritic β-amyloid plaques. Neuritic plaques virtually always contain these microglia, and in greater numbers. Among neuritic plaques, these activated microglia are more numerous in earlier stages (*i. e.*, those with diffuse rather than dense cores of β-amyloid). The putative end-stage, "burned out" plaques, consisting of dense β-amyloid cores without associated neuritic elements, are devoid of microglia. This pattern of association of activated microglia with different plaque types suggests a waxing and waning of an inflammation-like process in which microglial activation and IL-1 play an important role in plaque progression.

The amyloid plaques of Alzheimer's disease also contain activated astrocytes that overexpress the neurite growth-promoting (Kligman and Marshak

1985) cytokine S100β (Griffin et al. 1989; Marshak et al. 1992), an observation that suggests a role for this neuritotrophic cytokine in the induction of swollen, dystrophic neurites in β-amyloid plaques (Griffin et al. 1989; Sheng et al. 1994b). The pattern of distribution of these astrocytes among different plaque types parallels that of activated IL-1$^+$ microglia (Mrak et al. 1996), further supporting the idea of a waxing and waning inflammation-like process in plaque progression. Moreover, the numbers of activated astrocytes overexpressing S100β associated with individual neuritic plaques correlates with the quantity of dystrophic neurites overexpressing βAPP in individual neuritic plaques (Mrak et al. 1996), an observation that strongly supports the idea that S100β plays an important role in the induction of these neuritic elements.

Another potential contribution to plaque formation is IL-1 stimulation of astrocytes to synthesize and release proteins such as ApoE and α_1-antichymotrypsin (Das and Potter 1995). Both of these proteins are found in neuritic β-amyloid plaques in Alzheimer's disease (Sheng et al. 1996b; Strittmatter and Roses 1995; Abraham et al. 1988). The presence of ApoE may promote the aggregation of diffuse amyloid into fibrillar β-amyloid (Strittmatter and Roses 1995) and stimulate growth of processes (Crutcher et al. 1994).

The Cytokine Cycle in Neurodegeneration

A number of observations suggest that chronic IL-1 overexpression in brain, as is seen in Alzheimer's disease, sets in motion IL-1-driven cascades of cellular and molecular events with potentially damaging consequences. We first proposed (Griffin et al. 1989) that the downstream consequences of such IL-1 overexpression would favor IL-1-driven cascades and that these would lead to further neurodegeneration, not only in Alzheimer's disease and Down's syndrome but also in other conditions where glia are activated in response to neural insults. In addition to Alzheimer's disease and Down's syndrome, conditions in which chronic cerebral cortical overexpression of IL-1 is found include idiopathic epilepsy (Griffin et al. 1995b) and HIV infection (Tyor et al. 1992, Stanley et al. 1994). Acute overexpression of IL-1 is also seen following closed head injury (Griffin et al. 1994). It is of interest that all of these conditions also manifest astrocytic S100β overexpression (Griffin et al. 1989, 1995b; Marshak et al. 1992; Sheng et al. 1994b; Stanley et al. 1994; Gentleman SM et al. unpublished results), neuronal or neuritic βAPP overexpression (Gentleman et al. 1993; Cork et al. 1990; Mrak et al. 1996; Sheng et al. 1994a; Stanley et al. 1994; Griffin et al. 1994; MacKenzie and Miller 1994), and either increased risk for Alzheimer's disease or premature appearance of Alzheimer-type neuropathological changes (Wisniewski et al. 1985; MacKenzie and Miller 1994; Esiri et al. 1998; Mayeux et al. 1993; Mortimer et al. 1991).

These examples suggest that chronic overexpression of IL-1 in the brain may drive a cascade of cellular and molecular events, as is the case in the periphery. There is the potential for feedback amplification of these processes:

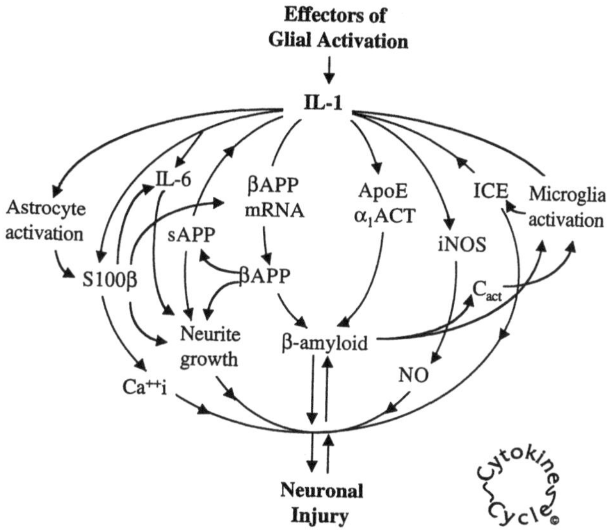

Fig. 1. Diagrammatic representation of the cytokine cycle, illustrating the way in which overexpression of IL-1 drives neurodegeneration and the way in which neurodegeneration drives further overexpression of IL-1, culminating in a vicious circle of neuronal loss and recharging of the cycle. Abbreviations: IL, interleukin; βAPP, β-amyloid precursor protein; sAPP, secreted βAPP; ApoE, apolipoprotein E; ICE, IL-1β converting enzyme; α_1ACT, α_1-anti-chymotrypsin; Ca^{++}i, intracellular free calcium concentration; NO, nitric oxide; iNOS, inducible NO synthase; C_{act}, complement activation

the potentially neurodegenerative consequences of such overexpression could produce neuronal cell injury, resulting in further microglial activation with further overexpression of IL-1. The result and consequence of these effects are a self-propagating *cytokine cycle* (Griffin et al. 1998a), in which chronic overexpression of IL-1: 1) stimulates astrocytes to become activated and to overexpress S100β ApoE, α_1-antichymotrypsin, and other Alzheimer-related proteins; 2) stimulates neurons to overexpress βAPP; and 3) provides feedback stimulation to microglia to become activated and further overexpress IL-1 (Fig. 1).

Potentially neurodegenerative consequences arise from each of these actions (Mrak et al. 1995). Chronic S100β overexpression by astrocytes leads to potentially toxic increases in neuronal intracellular free calcium concentrations (Barger and Van Eldik 1992) and to astrocytic synthesis and release of potentially toxic nitric oxide (Hu et al. 1996). IL-1-driven neuronal overexpression and processing of βAPP may increase release of neurotoxic (Mattson et al. 1992) β-amyloid and may indirectly promote β-amyloid plaque formation. β-Amyloid has also been shown to activate the classical complement pathway (Rogers et al. 1992), thus providing another potential mechanism for cellular injury, with consequent microglial activation for further feedback amplification of this cytokine cycle. Another consequence of IL-1-driven overexpression and increased processing of neuronal βAPP is the generation of non-amyloidogenic secreted βAPP fragments (sAPP). These sAPPs can activate microglia and promote microglial production of IL-1 and inducible nitric oxide synthase (Barger and Harmon 1997). Microglial activation and further IL-1 overexpression resulting from these several feedback mechanisms perpetuate the neurodegenerative process and thus propagate the cycle.

Role of Glial Cytokines in the Propagation of Neurodegeneration

A characteristic of Alzheimer's disease is progression of pathological changes across cerebral cortical regions. Although variations are seen in this progression, the earliest and severest involvement is generally found in the temporal lobes, with later involvement of the parietal, frontal, and occipital lobes, and little or no involvement of extracerebral regions such as the cerebellum. Both activated microglia, overexpressing IL-1, and activated astrocytes, overexpressing S100β, show patterns of distribution across brain regions in Alzheimer's disease that correlate with these generally recognized patterns of regional susceptibility (Sheng et al. 1995; Van Eldik and Griffin 1994). Moreover, local neuronal damage or compromise results in remote (axonal target area) activation of glia, overexpression of glial cytokines, and neuronal and neuritic overexpression of βAPP (Ito et al. 1997). We envision that such remote responses to local neuronal injury constitute the mechanistic basis for the spread of Alzheimer lesions across brain regions. Chronic glial activation arising at such remote sites may promote additional neuronal injury, βAPP overexpression, and neuritic alterations, setting in motion anew the cytokine cycle and propagating the pathological changes of Alzheimer's disease.

References

Abraham CR, Selkoe DJ, Potter H (1988) Immunochemical identification of the serine protease inhibitor alpha 1-antichymotrypsin in the brain amyloid deposits of Alzheimer's disease. Cell 52:487–501.

Andersen K, Launer LJ, Ott A, Hoes AW, Breteler MM, Hofman A (1995) Do nonsteroidal anti-inflammatory drugs decrease the risk for Alzheimer's disease? The Rotterdam Study. Neurology 45:1441–1445

Barger SW, Van Eldik LJ (1992) S100β stimulates calcium fluxes in glial and neuronal cells. J Biol Chem 267:9689–9694

Barger SW, Harmon AD (1997) Microglial activation by Alzheimer amyloid precursor protein and modulation by apolipoprotein E. Nature 388:878–881

Breitner JC, Gau BA, Welsh KA, Plassman BL, McDonald WM, Helms MJ, Anthony JC (1994) Inverse association of anti-inflammatory treatments and Alzheimer's disease: initial results of a co-twin control study. Neurology 44:227–232

Breitner JCS, Welsh KA, Helms MJ, Gaskell PC, Gau BA, Roses AD, Pericak-Vance MA, Saunders AM (1995) Delayed onset of Alzheimer's disease with nonsteroidal anti-inflammatory and histamine H2 blocking drugs. Neurobiol Aging 16:523–530

Brenneman DE, Page SW, Schultzberg M, Thomas FS, Zelazowski P, Burnet P, Avidor R, Sternberg EM (1992) A decomposition product of a contaminant implicated in L-tryptophan eosinophilia

myalgia syndrome affects spinal cord neuronal cell death and survival through stereospecific, maturation and partly interleukin-1-dependent mechanisms. J Pharmacol Exp Ther 266:1029–1035
Buxbaum JD, Oishi M, Chen HI, Pinkas-Kramarski R, Jaffe EA, Gandy SE, Greengard P (1992) Cholinergic agonists and interleukin 1 regulate processing and secretion of the Alzheimer β/A4 amyloid protein precursor. Proc Natl Acad Sci USA 89:10075 – 10078
Cacabelos R, Alvarez XA, Fernandez-Novoa L, Franco A, Mangues R, Pellicer A, Nishimura T (1994) Brain interleukin-1 beta in Alzheimer's disease and vascular dementia. Met Findings Exp Clin Pharm 16:141–151
Cork LC, Masters C, Beyreuther K, Price DL (1990) Development of senile plaques: relationships of neuronal abnormalities and amyloid deposits. Am J Pathol 137:1383–1392
Crutcher KA, Clay MA, Scott SA, Tian X, Tolar M, Harmony JA (1994) Neurite degeneration elicited by apolipoprotein E peptides. Exp Neurol 130:120–126
Das S, Potter H (1995) Expression of the Alzheimer amyloid-promoting factors α_1-antichymotrypsin and apolipoprotein E is induced in astrocytes by IL-1. Neuron 14:447–456
Dinarello CA, Wolff SM (1993) The role of interleukin-1 in disease. N Engl J Med 328:106-113
Esiri MM, Biddolph SC, Morris CS (1998) Prevalence of Alzheimer plaques in AIDS. J Neurol Neurosurg Psychiatry 65:29–33.
Forloni G, Demicheli F, Giorgi S, Bendotti C, Angeretti N (1992) Expression of amyloid precursor protein mRNAs in endothelial, neuronal and glial cells: modulation by interleukin-1. Brain Res Mol Brain Res 16:128–134
Ganter S, Northoff H, Mannel D, Gebicke-Harter PJ (1992) Growth control of cultured microglia. J Neurosci Res 33:218–230
Gentleman SM, Nash MJ, Sweeting CJ, Graham DI, Roberts GW (1993) β-Amyloid precursor protein (β-APP) as a marker for axonal injury after head injury. Neurosci Lett 160:139–144
Giulian D, Lachman LB (1985) Interleukin-1 stimulation of astroglial proliferation after brain injury. Science 228:497–499
Giulian D, Woodward J, Young DG, Krebs JF, Lachman LB (1988) Interleukin-1 injected into mammalian brain stimulates astrogliosis and neovascularization J Neurosci 8:2485–2490
Goldgaber D, Harris HW, Hla T, Maciag T, Donnelly RG, Jacobsen JS, Vitek MP, Gajdusek DC (1989) Interleukin 1 regulates synthesis of amyloid beta-protein precursor mRNA in human endothelial cells. Proc Natl Acad Sci USA 86:7606–7610
Griffin WST, Stanley LC, Ling C, White L, MacLeod V, Perrot LJ, White C III, Araoz C (1989) Brain interleukin 1 and S-100 immunoreactivity are elevated in Down syndrome and Alzheimer disease. Proc Natl Acad Sci USA 86:7611–7615
Griffin WST, Sheng JG, Gentleman SM, Graham DI, Mrak RE, Roberts GW (1994) Microglial interleukin-1α expression in human head injury: correlations with neuronal and neuritic β-amyloid precursor protein expression. Neurosci Lett 176:133–136
Griffin WST, Sheng JG, Roberts GW, Mrak RE (1995a) Interleukin-1 expression in different plaque types in Alzheimer's disease: significance in plaque evolution. J Neuropathol Exp Neurol 54:276–281
Griffin WST, Yeralan O, Boop F, Rovnaghi CR, Sheng JG, Mrak RE, Van Eldik LC (1995b) Overexpression of the neurotrophic cytokine S100β in human temporal lobe epilepsy. J Neurochem 65:228–233
Griffin WST, Sheng JG, Royston MC, Gentleman SM, Graham DI, Roberts GW, Mrak RE (1998a) Glial-neuronal interactions in Alzheimer's disease: the potential role of a 'cytokine cycle' in disease progression. Brain Pathol 8:65–72
Griffin WST, Sheng JG, MacKenzie JE, Royston MC, Gentleman SM, Brumback RA, Cork LC, Del Bigio MR, Roberts GW, Mrak RE (1998b) Life-long overexpression of S100β in Down's syndrome: Implications for Alzheimer pathogenesis. Neurobiol Aging 19:401–405
Hetier E, Ayala J, Denefle P, Bousseau A, Rouget P, Mallat M, Prochiantz A (1988) Brain macrophages synthesize interleukin-1 and interleukin-1 mRNAs *in vitro*. J Neurosci Res 21:391–397
Hu J, Castets F, Guevara JL, Van Eldik LJ (1996) S100β stimulates inducible nitric oxide synthase activity and mRNA levels in rat cortical astrocytes. J Biol Chem 271:2543–2547
Ito K, Ishikawa Y, Skinner RD, Mrak RE, Morrison-Bogorad M, Mukawa J, Griffin WST (1997) Lesioning of the inferior olive using a ventral surgical approach. Characterization of temporal

and spatial astrocytic responses at the lesion site and in cerebellum. Mol Chem Neuropathol 31:245–264

Kligman D, Marshak DR (1985) Purification and characterization of a neurite extension factor from bovine brain. Proc Natl Acad Sci USA 82:7136–7139

Lee SC, Liu W, Dickson DW, Brosnan CF, Berman JW (1993) Cytokine production by human fetal microglia and astrocytes. Differential induction by lipopolysaccharide and IL-1 beta. J Immunol 150:2659–2667

Marshak DR, Pesce SA, Stanley LC, Griffin WST (1992) Increased S100β neurotrophic activity in Alzheimer's disease temporal lobe. Neurobiol Aging 13:1–7

Mattson MP, Cheng B, Davis D, Bryant K, Lieberburg I, Rydel RE (1992) β-Amyloid peptides destabilize calcium homeostasis and render human cortical neurons vulnerable to excitotoxicity. J Neurosci 12:376–389

MacKenzie IR, Miller LA (1994) Senile plaques in temporal lobe epilepsy. Acta Neuropathol (Berl) 87:504–510

Mayeux R, Ottman R, Tang MX, Noboa-Bauza L, Marder K, Gurland B, Stern Y (1993) Genetic susceptibility and head injury as risk factors for Alzheimer's disease among community-dwelling elderly persons and their first-degree relatives. Ann Neurol 33:494–501

MacKenzie JE, Gentleman SM, Roberts GW, Graham DI, Royston MC (1994) Increased numbers of βAPP-immunoreactive neurons in the entorhinal cortex after head injury. NeuroReport 6:161–164

Mortimer JA, van Duijn CM, Chandra V, Fratiglioni L, Graves AB, Heyman A, Jorm AF, Kokmen E, Kondo K, Rocca WA, Shalat SL, Soininen H, Hofman A (1991) Head trauma as a risk factor for Alzheimer's disease: a collaborative re-analysis of case-control studies. EURODEM Risk Factors Research Group. Int J Epidemiol 20:S28–35

Mrak RE, Sheng JG, Griffin WST (1995) Glial cytokines in Alzheimer's disease: review and pathogenic implications. Human Pathol 26:816–823

Mrak RE, Sheng JG, Griffin WST (1996) Correlation of astrocytic S100β expression with dystrophic neurites in amyloid plaques of Alzheimer's disease. J Neuropathol Exp Neurol 55:273–279

Rich JB, Rasmusson DX, Folstein MF, Carson KA, Kawas C, Brandt J (1995) Nonsteroidal anti-inflammatory drugs in Alzheimer's disease. Neurology 45:51 – 55

Rogers J, Cooper NR, Webster S, Schultz J, McGeer PL, Styren SD, Civin WH, Brachova L, Bradt B, Ward P (1992) Complement activation by β-amyloid in Alzheimer disease. Proc Natl Acad Sci USA 89:10016–10020

Rothwell NJ (1991) Functions and mechanisms of interleukin 1 in the brain. TiPS 12:430–436

Rozemuller JM, Eikelenboom P, Stam FC, Beyreuther K, Masters CL (1989) A4 protein in Alzheimer's disease: primary and secondary cellular events in extracellular amyloid deposition. J Neuropathol Exp Neurol 48:674–691

Sebire G, Emilie D, Wallon C, Hery C, Devergne O, Delfraissy JF, Galanaud P, Tardieu M (1993) In vitro production of IL-6, IL-1 beta, and tumor necrosis factor-alpha by human embryonic microglial and neural cells. J Immunol 150:1517 – 1523

Selinfreund RH, Barger SW, Welsh MJ, Van Eldik LJ (1990) Antisense inhibition of glial S100β production results in alterations in cell morphology, cytoskeletal organization and cell proliferation. J Cell Biol 111:2021–2028

Selinfreund PH, Barger SW, Pledger WJ, Van Eldik LJ (1991) Neurotrophic protein S100β stimulates glial cell proliferation. Proc Natl Acad Sci USA 88:3554–3558

Sheng JG, Boop FA, Mrak RE, Griffin WST (1994a) Increased neuronal β-amyloid precursor protein expression in human temporal lobe epilepsy: Association with interleukin-1 alpha immunoreactivity. J Neurochem 63:1872–1879

Sheng JG, Mrak RE, Griffin WST (1994b) S100β protein expression in Alzheimer's disease: potential role in the pathogenesis of neuritic plaques. J Neurosci Res 39:398–403

Sheng JG, Mrak RE, Griffin WST (1995) Interleukin-1α expression in brain regions in Alzheimer's disease: Correlation with neuritic plaque distribution. Neuropathol Appl Neurobiol 21:290–301

Sheng JG, Ito K, Skinner RD, Mrak RE, Rovnaghi CR, Van Eldik LJ, Griffin WST (1996a) *In vivo* and *in vitro* evidence supporting a role for the inflammatory cytokine interleukin-1 as a driving force in Alzheimer pathogenesis. Neurobiol Aging 17:761-766

Sheng FG, Mrak RE, Griffin WST (1996b) Apolipoprotein E distribution among different plaque types in Alzheimer's disease: implications for its role in plaque progression. Neuropathol Appl Neurobiol 22:334-s-341

Sheng JG, Mrak RE, Griffin WST (1997) Glial-neuronal interactions in Alzheimer's disease: Progressive association of IL-1α^+ microglia and S100β^+ astrocytes with neurofibrillary tangle stages. J Neuropathol Exp Neurol 56:285 – 290

Stanley LC, Mrak RE, Woody RC, Perrot LJ, Zhang SX, Marshak DR, Nelson SJ, Griffin WST (1994) Glial cytokines as neuropathogenic factors in HIV infection: Pathogenic similarities to Alzheimer's disease. J Neuropathol Exp Neurol 53:231–238.

Strijbos PJ, Rothwell NJ (1995) Interleukin-1β attenuates excitatory amino acid-induced neurodegeneration *in vitro:* involvement of nerve growth factor. J Neurosci 15:3468–3474

Strittmatter WJ, Roses AD (1995) Apolipoprotein E: emerging story in the pathogenesis of Alzheimer's disease. The Neuroscientist 1:298–306

Tyor WR, Glass JD, Griffin JW, Becker PS, McArthur JC, Bezman L, Griffin DE (1992) Cytokine expression in the brain during the acquired immunodeficiency syndrome. Ann Neurol 31:349–360

Wisniewski KE, Wisniewski HM, Wen GY (1985) Occurrence of neuropathological changes and dementia of Alzheimer's disease in Down's syndrome, Ann Neurol 17:278–282

Van Eldik LJ, Griffin WST (1994) S100β expression in Alzheimer's disease: Relation to neuropathology in brain regions. Biochim Biophys Acta 39:398–404

Contributions of the Glial Injury Response to the Multifactorial Pathogenesis of Alzheimer's Disease

L. Mucke, M. Buttini, R.W. Mahley, R.E. Pitas, J. Raber, and T. Wyss-Coray

Introduction

Alzheimer's disease (AD) is the leading cause of cognitive deficits in the elderly. As longevity of many populations has increased, so has the number of people suffering from this devastating illness (Ott et al. 1998).

The risk of developing AD appears to depend on a number of different genetic and epigenetic factors. Apolipoprotein E4 (apoE4) is a major determinant of cognitive impairments, risk of developing "sporadic" AD, and poor outcome after head injury or stroke, while mutations in genes encoding human amyloid β protein precursor (hAPP) or presenilins cause autosomal dominant forms of familial AD (FAD; Strittmatter and Roses 1996; Weisgraber and Mahley 1996; Farrer et al. 1997; Lendon et al. 1997; Selkoe 1997). As outlined below, evidence is accumulating that cytokines can form pathogenetic links between genetic predispositions and epigenetic factors such as age and head injury (Fig. 1).

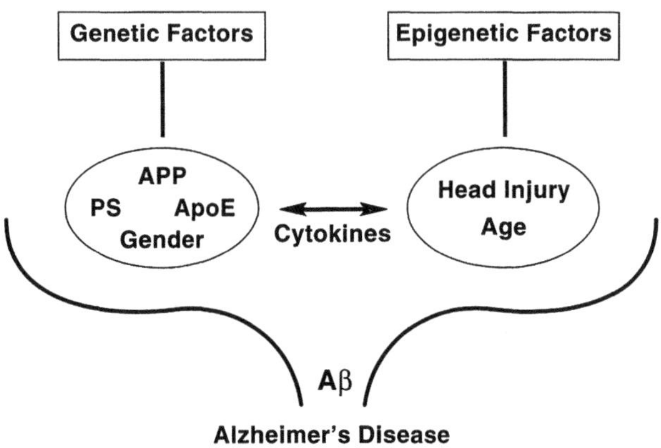

Fig. 1. Multifactorial pathogenesis of AD. Based on current knowledge, it is likely that distinct constellations of genetic and epigenetic factors result in the development of AD in different patients. Glia-derived injury response factors such as cytokines could form critical links between these components. To assess the potential value of anti-inflammatory and anti-amyloid drug treatments, future research will have to determine whether inflammatory mediators and Aβ are sufficient or necessary for the completion of this pathogenetic cascade. ApoE, apoliprotein E; APP, amyloid precursor protein; PS, presenilin

Patterson/Kordon/Christen (Eds.)
Neuro-Immune Interactions in Neurologic and Psychiatric Disorders

Contribution of Inflammatory Mediators to Neurological Diseases

While the primary causes and proximal triggers of neural injury vary greatly among different neurological diseases, diverse pathogenetic cascades can converge onto a limited number of final common pathways that culminate in neuronal dysfunction or death. CNS inflammation represents an example of such a final common pathogenetic pathway. Key components of these pathways are attractive therapeutic targets, as their pharmacological manipulation may allow treatment of diverse diseases. However, this strategy is complicated by the fact that these mediators often fulfill many different functions, not only in abnormal but also in normal CNS biology, and that some of their physiological functions may be of fundamental importance.

Evidence has been accumulating that cytokines and other inflammatory mediators may contribute to various neurological diseases, including AD (McGeer and McGeer 1995). Although blood-derived immune cells can infiltrate the CNS, particularly when the blood-brain barrier has been compromised, their invasion of the CNS is frequently delayed, and their activity in the CNS parenchyma is limited compared with immune responses targeted to other organs. Indeed, in most chronic neurodegenerative diseases, classical inflammatory infiltration of the CNS by hematogenous cells is typically absent. The CNS possesses an elaborate network of resident cells (astrocytes and microglia) that respond quickly and vigorously to diverse neural injuries. Activation of these cells (gliosis) results in morphological alterations as well as in dramatic qualitative and quantitative changes in the vast array of molecules that these cells synthesize (Eddleston and Mucke 1993; Mucke and Eddleston 1993). These injury-response factors are probably produced pri-

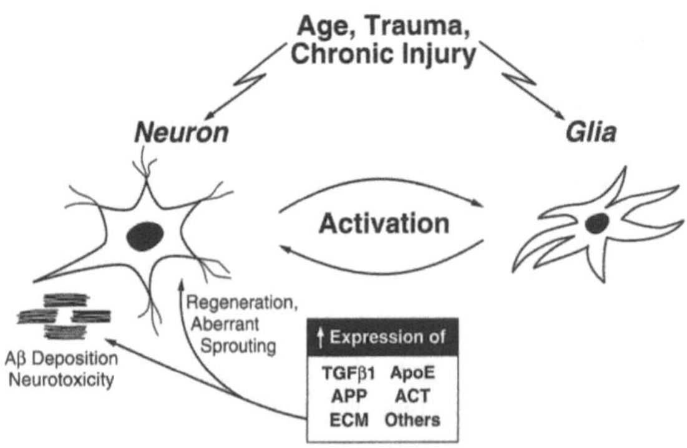

Fig. 2. Is glial activation beneficial or detrimental? Brain injury results in the increased production of several molecules that, depending on their concentration and time course of expression, can either fulfill neurotrophic/neuroprotective functions or contribute to amyloidogenesis and neurodegeneration. This dichotomy of effects raises concerns about the therapeutic value of suppressing glial activation in general and suggests that regulation of specific molecular pathways may be more beneficial. ACT, α_1-antichymotrypsin; ApoE, apolipoprotein E; APP, amyloid precursor protein; ECM, extracellular matrix components; TGF-β1, transforming growth factor-β1.

marily to protect the CNS and to help in neuronal regeneration after injury. However, inflammation and injury responses can also contribute to disease progression, particularly when they become chronic (Fig. 2).

Transforming growth factor – β1 (TGF-β1) is a multifunctional cytokine that plays key roles in neuroprotection, wound repair, and immune responses (Finch et al. 1993; Prehn et al. 1994, 1996; Flanders et al. 1998). Several neurological diseases, including AD, HIV encephalitis, and communicating hydrocephalus, are associated with increased expression of TGF-β1 in the CNS (Wahl et al. 1991; Finch et al. 1993; van der Wal et al. 1993; Chao et al. 1994; Wyss-Coray et al. 1995). Because TGF-β1 has been shown in cell culture models to fulfill multiple functions, some of which could have opposing effects on disease progression, it is difficult to predict which of its many effects prevails in any given pathophysiological scenario *in vivo*. Therefore, we used transgenic mice overexpressing TGF-β1 from astrocytes to assess the contributions of this cytokine to different CNS diseases (Wyss-Coray et al. 1995, 1997 a, b).

When expressed at high levels either early in development or acutely in adulthood, TGF-β1 induces a vast overexpression of extracellular matrix components in the brain. This overexpression appears to interfere with CSF turnover and results in the development of communicating hydrocephalus (Wyss-Coray et al. 1995). Low levels of TGF-β1 expression in the CNS *per se* did not induce immunopathological CNS alterations but functioned as an important pathogenic co-factor in inflammatory demyelinating disease. Although TGF-β1 suppresses experimental autoimmune encephalomyelitis (EAE) when injected systemically (Racke et al. 1993), local overexpression of TGF-β1 in the spinal cord enhanced the recruitment of immune cells into the CNS parenchyma and exacerbated the clinical course of EAE (Wyss-Coray et al. 1997 a).

Interestingly, low-level chronic astroglial overexpression of TGF-β1 resulted in the development of cerebrovascular amyloid deposits in aging mice and, in the presence of high Aβ levels, accelerated the deposition of amyloid in cerebral blood vessels (Wyss-Coray et al. 1997 b). As outlined in greater detail below, our molecular and neuropathological analyses of *postmortem* brain tissues indicate that such TGF-β1 effects may also contribute to the development of cerebral amyloid angiopathy (CAA) in humans with AD.

Despite the increasing evidence that cerebral Aβ deposition plays an important role in AD pathogenesis, the factors that initiate or promote Aβ deposition in sporadic AD remain largely unknown. Amyloidogenic mutations in APP and presenilin genes linked to autosomal dominant FAD are not seen in sporadic AD cases, and neither is the widespread overexpression of APP that is associated with AD-like CNS alterations in humans with Down's syndrome and in hAPP transgenic mouse models. Therefore, the identification of molecules promoting Aβ deposition in the brain has become a major quest in laboratories around the world.

Our studies indicate that TGF-β1 plays an important role in the deposition of amyloid in cerebral blood vessels (Wyss-Coray et al. 1997 b). This finding is interesting for a number of reasons. Evidence has been accumulat-

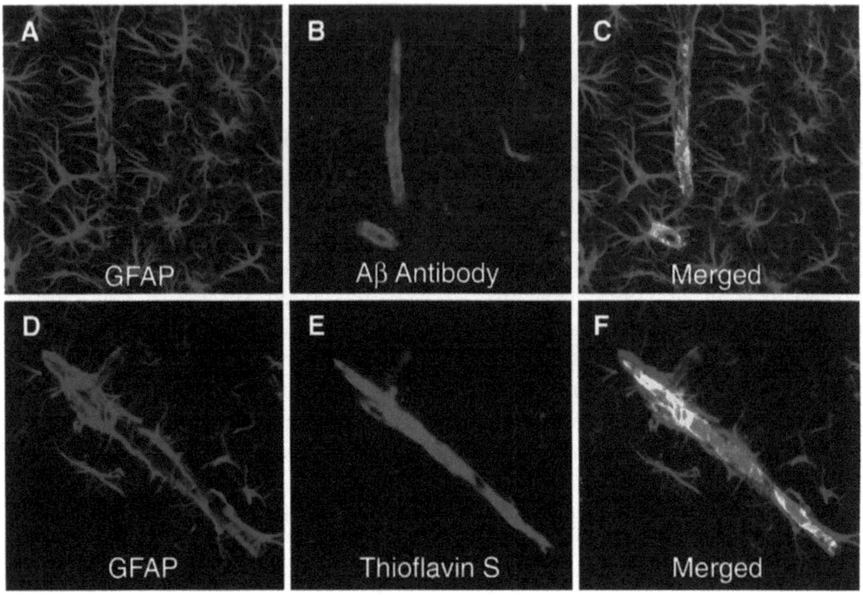

Fig. 3 A–F. Cerebrovascular amyloidosis in TGF-β1+hAPP bigenic mice. Perivascular astrocytosis and cerebrovascular amyloid deposits in 10-week-old bigenic mice (PDGF-hAPP line H6 x GFAP-TGF-β1) were revealed by confocal microscopy. Brain sections were double-labeled with antibodies against Aβ and against the astroglial marker, glial fibrillary acidic protein (GFAP; **A–C**) or with thioflavin S staining and antibodies against GFAP (**D–F**) as indicated. Confocal images were collected via FITC and Cy5 channels. Digital merge of images from the same section revealed a close association between perivascular astrocytes and vascular amyloid deposits

ing that cerebrovascular alterations may contribute significantly to AD pathogenesis (Roses and Saunders 1997). Although it has long been recognized that CAA, a condition resulting from abnormal deposition of β amyloid in cerebral blood vessels, is frequently associated with AD, the cause of CAA has remained a complete mystery. This is unfortunate not only because of its potential role in AD but also because, in its most severe form, CAA constitutes an important cause of devastating brain hemorrhages in the elderly (Desmond 1996; Green and Vonsattel 1997).

Our combined molecular and neuropathological analysis of *postmortem* brain tissues from humans with AD and nondemented controls revealed a strong positive correlation between levels of TGF-β1 expression and extent of cerebrovascular amyloid deposition (Wyss-Coray et al. 1997b). Because these correlation studies, by themselves, did not prove a cause-effect relationship between TGF-β1 and cerebrovascular amyloidogenesis, we complemented our analysis of human tissues with an investigation of related transgenic models.

Aged transgenic mice overexpressing TGF-β1 in astrocytes developed thioflavin S-positive and Aβ-immunoreactive deposits in cerebral blood vessels and meninges (Wyss-Coray et al. 1997b). Double transgenic mice in which TGF-β1 was co-expressed with FAD-mutant human APP (as a source

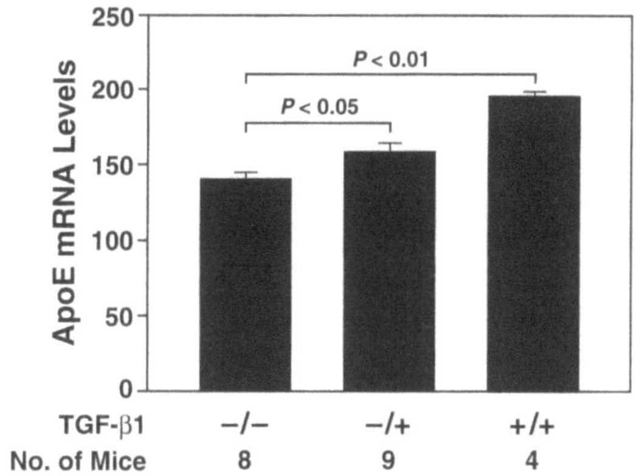

Fig. 4. TGF-β1 increases cerebral apoE expression. ApoE mRNA levels in non-transgenic controls (–/–) and in mice heterozygous (–/+) or homozygous (+/+) for the GFAP-TGF-β1 transgene were determined by RNase protection assay and quantitation of radioactive signals with a phosphorimager. Data represent means ± SEM. *P* values were determined by Dunnett's posthoc test

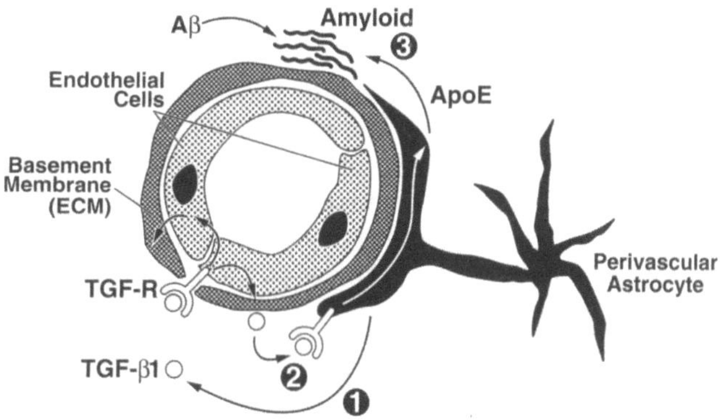

Fig. 5. Diagram depicting possible interactions between TGF-β1 and (peri)vascular cells. Injury-induced upmodulation of TGF-β1 secretion by astrocytes (1) may result in increased vascular production of extracellular matrix components (ECM) and basement membrane, as well as in autocrine stimulation of TGF-β1 production (2). This process may in turn lead to a chronic perivascular astrocytosis, with TGF-β1-induced increases in astroglial secretion of apoE (3). Amyloidogenic interactions of apoE with Aβ may then close the vicious cycle of astroglial activation and vascular amyloid deposition

of Aβ) showed cerebrovascular deposition of human Aβ by 10 weeks of age (Fig. 3), whereas no such deposits were detected in singly transgenic hAPP or TGF-β1 littermates (Wyss-Coray et al. 1997b). These results indicate that TGF-β1 can effectively initiate or accelerate the deposition of Aβ in cerebral blood vessels. Ongoing studies focus on the dissection of the molecular processes that underlie these amyloidogenic TGF-β1 effects.

Additional support for an amyloidogenic role of TGF-β1 comes from studies in rats in which infusion of Aβ produced cerebral Aβ-immunoreactive deposits only if the animals also received a ventricular injection of recombinant TGF-β1 (Frautschy et al. 1996). Furthermore, at the sixth International Conference on Alzheimer's Disease and Related Disorders (Amsterdam, 1998), recombinant TGF-β1 was shown to promote the accumulation or deposition of Aβ in cortical slice cultures (Harris-White et al. 1998), cultured canine cerebral blood vessels (Mazur-Kolecka et al. 1998), and cultured smooth muscle cells (Prior et al. 1998).

Recently, we found in transgenic mice that TGF-β1 increases brain expression of the apolipoprotein E *(Apoe)* gene (Fig. 4) and that genetic elimination of apoE decreases cerebrovascular amyloid deposition after brain injury (Wyss-Coray et al. 1998). Human AD cases also showed a strong positive correlation between TGF-β1 and apoE mRNA levels, and addition of recombinant TGF-β1 to primary cultures of mouse astrocytes stimulated their secretion of apoE (Wyss-Coray et al. 1998). These findings suggest that TGF-β1 may contribute to the development of CAA by increasing the expression of pathological molecular chaperones and that apoE may function as an essential mediator of TGF-β1-induced cerebral amyloidosis (Fig. 5).

Differential CNS Effects of Human ApoE Isoforms

Evidence is increasing that apoE plays a central role in the development of sporadic AD as well as in the response of the CNS to neuronal injury. Apolipoprotein E4, which differs from the more common apoE3 by only a single amino acid, significantly increases the risk of developing AD (Strittmatter and Roses 1996; Weisgraber and Mahley 1996; Farrer et al. 1997). Although diverse hypotheses have been advanced to explain how apoE4 might increase AD risk, the effects of distinct human apoE isoforms on the CNS remain to be defined.

Among different organs, the brain is second only to the liver with respect to relative apoE levels. Although astrocytes are the main source of apoE in rodent brain, apoE immunoreactivity (Han et al. 1994; Bao et al. 1996; Metzger et al. 1996; Raber et al. 1998) and apoE mRNA (Xu et al. 1999) have been detected in neurons of human AD brains. In addition, human apoE isoforms show similar differences on neurite outgrowth in cell culture regardless of whether apoE is added to culture media, released from co-cultured astrocytes, or produced directly by transfected neurons (Fig. 6). We therefore used the neuron-specific enolase (NSE) promoter to express human apoE3 or apoE4 at comparable levels in neurons of mice lacking murine apoE (*Apoe*$^{-/-}$; Raber et al. 1998; Buttini et al. 1999).

First, we assessed spatial learning and memory, which is severely impaired in AD. NSE-apoE4 mice were significantly impaired in acquisition of a water maze task compared with both wild-type mice and Apoe$^{-/-}$ littermate controls lacking human apoE expression. This finding indicates that the expression of apoE4 is worse than having no apoE at all, consistent with a pathogenic gain of function. In contrast, NSE-apoE3 mice performed as well as or better than wild-type mice in the water maze task, suggesting that apoE3 can more than adequately compensate for the lack of murine apoE in the brain of Apoe$^{-/-}$ mice. Compared with wild-type controls, NSE-apoE4 and Apoe$^{-/-}$ mice, but not NSE-apoE3 mice, also showed significant impair-

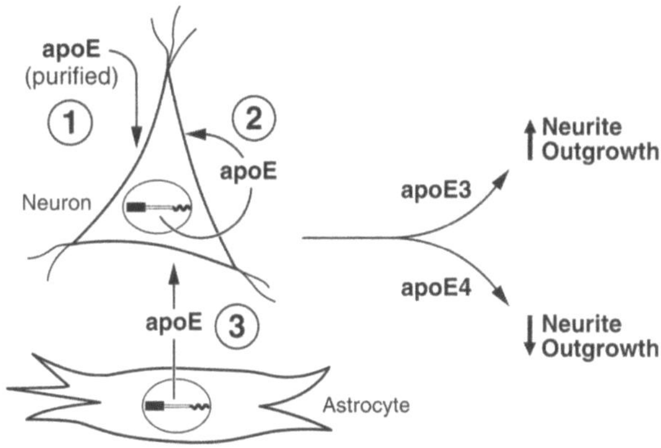

Fig. 6. ApoE isoform-specific effects on neurite outgrowth. As indicated, human apoE isoforms exert differential effects on neurite outgrowth by cultured neuronal cells, regardless of whether they are added to culture medium in purified form (Nathan et al. 1994; (1), expressed in neuronal cells via stable transfection (Bellosta et al. 1995; (2), or secreted from co-cultured astrocytes (Sun et al. 1998; (3)

ments in vertical exploratory behavior. All apoE4-induced behavioral deficits increased with age and were seen primarily in females. These findings demonstrate that distinct human apoE isoforms have differential effects on brain function *in vivo,* and that age and gender influence the susceptibility to apoE4-induced functional CNS impairments. They also underline that APOE genotype-gender interactions have to be considered in the design and assessment of AD treatments. Although great caution must be applied in extrapolating from findings in mice to complex human diseases, the results from the above analyses may provide guidance in the decision whether to pharmacologically inhibit detrimental apoE4 effects or to simulate apoE3 effects in human carriers of the APOE ε4 allele.

To begin to dissect the mechanisms by which apoE4 impairs cognitive functions, we asked whether apoE3 and apoE4 differ in their ability to protect neurons against age-related and excitotoxin-induced neurodegeneration. We first compared the baseline integrity of neuronal structures in the CNS of untreated mice at three to four months and at seven to nine months of age. In contrast to wild-type mice, *Apoe*$^{-/-}$ mice developed an age-dependent loss of presynaptic terminals and neuronal dendrites in the neocortex and hippocampus, two brain structures essential for cognitive function. Notably, expression of apoE3 effectively prevented this age-related neurodegeneration in transgenic *Apoe*$^{-/-}$ mice, whereas expression of apoE4 did not. Furthermore, following systemic injection of excitotoxins, NSE-apoE4 mice showed significantly more neurodegeneration than NSE-apoE3 mice or wild-type controls. The differential CNS effects of human apoE isoforms identified in NSE-apoE mice may relate closely to the increased susceptibility to AD observed in humans expressing apoE4.

Table 1. Findings obtained in different groups of *Apoe* knockout mice

Parameters assessed	Abnormal/decreased	Normal
Learning and memory		
Spatial learning	(Gordon et al. 1995; Masliah et al. 1997; Oitzl et al. 1997; Zhou et al. 1998)	(Anderson and Higgins 1997; Anderson et al. 1998; Raber et al. 1998)
Explorative behavior	(Raber et al. 1998)	
Long term potentiation (LTP)	(Krugers et al. 1997; Veinbergs et al. 1998)	(Anderson and Higgins 1997; Anderson et al. 1998)
Developmental milestones	(Gozes et al. 1997)	
Neuropathology		
Synaptophysin	(Masliah et al. 1995a,b, 1997; Buttini et al. 1999)	(Anderson and Higgins 1997; Anderson et al. 1998)
Microtubule associated protein (MAP2)	(Masliah et al. 1995b, 1997; Buttini et al. 1999)	(Anderson and Higgins 1997; Anderson et al. 1998)
Tubulin	(Masliah et al. 1995b)	
Phosphorylated neurofilaments	(Buttini et al. 1999)	(Mercken and Brion 1995)
Tau phosphorylation	(Genis et al. 1995)	
Postsynaptic receptor density	(Chapman and Michaelson 1998)	
Unmyelinated axons (PNS)	(Fullerton et al. 1998)	
Response to injury		
Entorhinal cortex lesion	(Masliah et al. 1995; Anderson and Higgins 1997; Anderson et al. 1998; Stone et al. 1998)	
Closed head injury	(Chen et al. 1997; Lomnitski et al. 1997)	
Excitotoxicity-induced neuronal injury		(Buttini et al. 1999)
Fimbria fornix lesion		(Fagan et al. 1998)
Remyelinization (PNS)		(Goodrum et al. 1995)
Middle cerebral artery occlusion	(Laskowitz et al. 1997)	
Neuropharmacology		
Choline acetyltransferase (ChAT)	(Gordon et al. 1995; Gozes et al. 1997)	(Anderson and Higgins 1997; Anderson et al. 1998; Fagan et al. 1998)
Vasoactive intestinal peptide (VIP)	(Gozes et al. 1997)	

While the differences we observed between NSE-apoE3 and NSE-apoE4 transgenic mice are robust, there is currently a puzzling amount of controversy in the field regarding the presence of CNS impairments in mice lacking apoE expression. Pertinent studies are summarized in Table 1. For example, in agreement with results obtained by Anderson et al. (1998), we found normal spatial learning in six-month-old *Apoe* knockout males, whereas others have reported mild (Gordon et al. 1995) or severe (Oitzl et al. 1997) spatial learning deficits in such mice. Yet, in contrast to Anderson et al. (1998), we have confirmed the age-dependent development of neurodegeneration in

Apoe knockout mice previously reported by Masliah and his colleagues (1995b). The reasons for these discrepancies are unclear. Differences in mouse strains and experimental methods (Gerlai 1996; Nelson 1997; Owen et al. 1997) obviously deserve consideration but do not appear to satisfactorily explain all differences observed. It is conceivable that specific dietary or other housing-related factors could increase age-related stresses on neurons and, thereby, help reveal the lack of neuroprotective apoE effects in some cohorts of aging $Apoe^{-/-}$ mice. Attempts to reliably resolve the apparent discrepancies among different cohorts of $Apoe^{-/-}$ mice will likely require concerted long-range efforts among different laboratories, since all groups of mice will have to be backcrossed onto the same genetic substrain and analyzed side-by-side with the same methodologies.

Transgenic Models to Assess ApoE-Aβ Interactions *In vivo* and to Predict the Effectiveness of ApoE-related Drug Treatments

While the increased expression of injury response factors by reactive glia likely fulfills neuroprotective functions, chronic overexpression of these factors may contribute to AD pathogenesis, particularly if the glial products are rendered deleterious by genetic alterations or pathogenic interactions with the Aβ peptide.

There is evidence from *in vitro* data, albeit controversial, that apoE3 and apoE4 could have differential effects on Aβ deposition, clearance, or both (see Weisgraber and Mahley 1996 and Aleshkov et al. 1997 for review). Elimination of murine apoE expression diminished amyloid plaque formation in the brains of APP transgenic mice but did not interfere with transgene expression or production of human Aβ, suggesting that murine apoE promotes the cerebral deposition of human Aβ *in vivo* (Bales et al. 1997). In humans, the APOE ε4 allele has been shown to increase the amount of Aβ deposited in plaques and cerebral blood vessels (see Hyman et al. 1996 for review).

Although Aβ has been shown to be neurotoxic in cell culture (Yankner et al. 1990; Pike et al. 1993; Lambert et al. 1998), a causal role for Aβ in neuronal degeneration *in vivo* remains speculative. A particularly controversial question is whether Aβ-induced neurotoxicity requires the formation of amyloid plaques (Cummings et al. 1996; Terry 1996; Bartoo et al. 1997; Davis and Chisholm 1997; Gomez-Isla et al. 1997; Irizarry et al. 1997a,b; Calhoun et al. 1998). Interestingly, recent evidence suggests that Aβ neurotoxicity could be induced by diffusible Aβ oligomers (Lambert et al. 1998) or by the intraneuronal accumulation of insoluble Aβ (Masliah et al. 1996; Turner et al. 1996). Cell culture studies (Miyata and Smith 1996) and the *in vivo* data described above suggest that apoE3 could protect neurons against Aβ-induced as well as other types of injury, whereas apoE4 may be deficient in this function or exert detrimental effects of its own. To further elucidate these complex interactions *in vivo*, we have generated doubly transgenic mice in which neuronal or astroglial expression of individual apoE isoforms is combined with neuro-

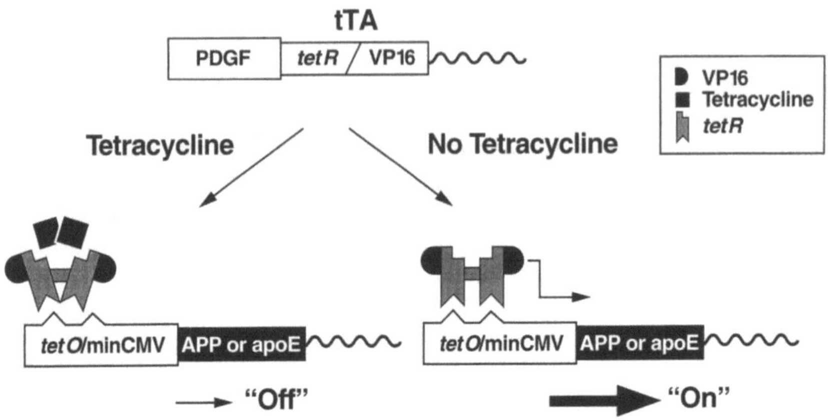

Fig. 7. Schematic outline of the regulatable promoter system. The tetracycline-regulatable promoter system was developed by Gossen and Bujard (1992) and has already been used successfully in transgenic models (Hennighausen et al. 1995; Mayford et al. 1996). In the scenario depicted above, the PDGF promoter directs expression of the transactivator (tTA), which consists of the tetR repressor from the tetracycline-resistance operon of *E. coli* and the eukaryotic activating domain of the herpes simplex virus protein VP16, the latter protein converts the repressor into a transcription activator. In the absence of tetracycline, tTA binds to the operator sites (tetO) linked to a minimal human cytomegalovirus (minCMV) promoter and activates transcription. In the presence of tetracycline or doxycycline, tTA dissociates from its binding sites, which turns off the minCMV-driven transcription unit. DNA sequences encoding FAD-mutant hAPP, apoE3, or apoE4 were placed downstream of the tetO/minCMV promoter. Various promoters are being used to target tTA expression to neurons, astrocytes, or microglia

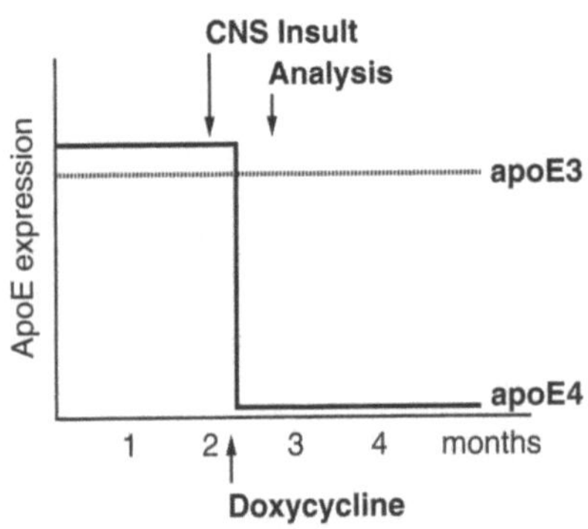

Fig. 8. Genetic simulation of anti-apoE4 drug effects in regulatable models. Diagram depicts how transgenic mice with regulatable apoE expression can be used to help predict whether reducing apoE4 activity in an APOE ε3/ε4 heterozygote might improve outcome after stroke, head injury, or neurosurgical interventions. Mice expressing apoE3 from a constitutive promotor and apoE4 from a regulatable promoter are challenged with a CNS injury. ApoE4 expression is suppressed with doxycycline at different time points and for different time periods after the insult. Behavioral, electrophysiological, and histopathological analyses are then used to compare the efficacy of the different "treatment regimens"

nal overexpression of FAD-mutant amyloidogenic APPs. The analysis of these models is in progress.

As outlined above, we have previously shown that the constitutive expression of FAD-mutant hAPP or apoE4 in brains of transgenic mice induces structural and functional CNS impairments that in several respects resemble AD (Games et al. 1995; Masliah et al. 1996; Wyss-Coray et al. 1997b; Raber et al. 1998). Recently, we have begun to use a regulatable promoter system (Fig. 7) to express FAD-mutant hAPP, apoE3, or apoE4 in brains of transgenic mice. We have established multiple lines of tetO-hAPP mice, as well as lines of mice transgenic for tetO-apoE3 or tetO-apoE4. Crosses between these lines and mice with cell-specific tTA expression have yielded double transgenic lines with regulatable expression of hAPP or apoE. These new models will help predict the therapeutic usefulness of manipulating individual pathogenic factors at baseline as well as in complex clinical situations (Fig. 8).

Acknowledgments. We thank G. Howard and S. Ordway for editorial assistance, J. Carroll and C. Goodfellow for help with graphics, and G. Costa for administrative assistance. This work was supported by a grant from the United States Public Health Services (AG11385 to L.M) and by a Cambridge NeuroScience/Gladstone collaborative research agreement.

References

Aleshkov S, Abraham CR, Zannis VI (1997) Interaction of nascent apoE2, apoE3, and ApoE4 isoforms expressed in mammalian cells with amyloid peptide β (1–40). Relevance to Alzheimer's disease. Biochemistry 36:10571–10580

Anderson R, Higgins GA (1997) Absence of central cholinergic deficits in apoE knockout mice. Psychopharmacology 132:135–144

Anderson R, Barnes JC, Bliss TVP, Cain DP, Cambon K, Davies HA, Errington ML, Fellows LA, Gray RA, Hoh T, Stewart M, Large CH, Higgins GA (1998) Behavioural, physiological and morphological analysis of a line of apolipoprotein E knockout mouse. Neuroscience 85:93–110

Bales KR, Verina R, Dodel RC, Du R, Altstiel L, Bender B, Hyslop P, Johnstone EM, Little SP, Cummins DJ, Piccardo P, Ghetti B, Paul SM (1997) Lack of apolipoprotein E dramatically reduces amyloid β-peptide deposition. Nat Genet 17:263–264

Bao F, Arai H, Matsushita S, Higuchi S, Sasaki H (1996) Expression of apolipoprotein E in normal and diverse neurodegenerative disease brain. NeuroReport 7:1733–1739

Bartoo GT, Nochlin D, Chang D, Kim Y, Sumi SM (1997) The mean Aβ load in the hippocampus correlates with duration and severity of dementia in subgroups of Alzheimer disease. J Neuropathol Exp Neurol 56:531–540

Bellosta S, Nathan BP, Orth M, Dong L-M, Mahley RW, Pitas RE (1995) Stable expression and secretion of apolipoproteins E3 and E4 in mouse neuroblastoma cells produces differential effects on neurite outgrowth. J Biol Chem 270:27063–27071

Buttini M, Orth B, Bellosta S, Akeefe H, Pitas RE, Wyss-Coray T, Mucke L, Mahley RW (1999) Expression of human apolipoprotein E3 or E4 in the brains of *Apoe*$^{-/-}$ mice: Isoform-specific effects on neurodegeneration. J Neurosci 19:4867–4880

Calhoun ME, Wiederhold KH, Abramowski D, Phinney AL, Probst A, Sturchler-Pierrat C, Staufenbiel M, Sommer B, Jucker M (1998) Neuron loss in APP transgenic mice. Nature 395:755–756

Chao CC, Hu S, Frey WH, II, Ala TA, Tourtellotte WW, Peterson PK (1994) Transforming growth factor β in Alzheimer's disease. Clin Diag Lab Immunol 1:109–110

Chapman S, Michaelson DM (1998) Specific neurochemical derangements of brain projecting neurons in apolipoprotein E-deficient mice. J Neurochem 70:708–714

Chen Y, Lomnitski L, Michaelson DM, Shohami E (1997) Motor and cognitive deficits in apolipoprotein E-deficient mice after closed head surgery. Neuroscience 80:1255–1262

Cummings BJ, Pike CJ, Shankle R, Cotman CW (1996) β-amyloid deposition and other measures of neuropathology predict cognitive status in Alzheimer's disease. Neurobiol Aging 17:921–933

Davis JN, Chisholm JC (1997) The 'amyloid cascade hypothesis' of AD: Decoy or real McCoy? Trends Neurosci 20:558–559

Desmond DW (1996) Vascular dementia: A construct in evolution. Cerebrovasc Brain Metab Rev 8:296–325

Eddleston MP, Mucke L (1993) Molecular profile of reactive astrocytes – implications for their role in neurologic disease. Neuroscience 54:15–36

Fagan AM, Murphy BA, Patel SN, Kilbridge JF, Mobley WC, Bu GJ, Holtzman DM (1998) Evidence for normal aging of the septo-hippocampal cholinergic system in apoE (-/-) mice but impaired clearance of axonal degeneration products following injury. Exp Neurol 151:314–325

Farrer LA, Cupples LA, Haines JL, Hyman B, Kukull WA, Mayeux R, Myers RH, Pericak-Vance MA, Risch N, van Duijn CM (1997) Effects of age, sex, and ethnicity on the association between apolipoprotein E genotype and Alzheimer disease. A meta-analysis. JAMA 278:1349–1356

Finch CE, Laping NJ, Morgan TE, Nichols NR, Pasinetti GM (1993) TGF-β1 is an organizer of responses to neurodegeneration. J Cell Biochem 53:314–322

Flanders KC, Ren RF, Lippa CF (1998) Transforming growth factor-βS in neurodegenerative disease. Prog Neurobiol 54:71–85

Frautschy SA, Yang F, Calderón L, Cole GM (1996) Rodent models of Alzheimer's disease: Rat Aβ infusion approaches to amyloid deposits. Neurobiol Aging 17:311–321

Fullerton SM, Strittmatter WJ, Matthew WD (1998) Peripheral sensory nerve defects in apolipoprotein E knockout mice. Exp Neurol 153:156–163

Games D, Adams D, Alessandrini R, Barbour R, Berthelette P, Blackwell C, Carr T, Clemens J, Donaldson T, Gillespie F, Guido T, Hagopian S, Johnson-Wood K, Khan K, Lee M, Leibowitz P, Lieberburg I, Little S, Masliah E, McConlogue L, Montoya-Zavala M, Mucke L, Paganini L, Penniman E, Power M, Schenk D, Seubert P, Snyder B, Soriano F, Tan H, Vitale J, Wadsworth S, Wolozin B, Zhao J (1995) Alzheimer-type neuropathology in transgenic mice overexpressing V717F β-amyloid precursor protein. Nature 373:523–527

Genis I, Gordon I, Sehayek E, Michaelson DM (1995) Phosphorylation of tau in apolipoprotein E-deficient mice. Neurosci Lett 199:5–8

Gerlai R (1996) Gene-targeting studies of mammalian behavior: Is it the mutation or the background genotype? Trends Neurosci 19:177–181

Gomez-Isla T, Hollister R, West H, Mui S, Growdon JH, Petersen RC, Parisi JE, Hyman BT (1997) Neuronal loss correlates with but exceeds neurofibrillary tangles in Alzheimer's disease. Ann Neurol 41:17–24

Goodrum JF, Bouldin TW, Zhang SH, Maeda N, Popko B (1995) Nerve regeneration and cholestrol reutilization occur in the absence of apolipoproteins E and A-I in Mice. J Neurochem 64:408–416

Gordon I, Grauer E, Genis I, Sehayek E, Michaelson DM (1995) Memory deficits and cholinergic impairments in apolipoprotein E-deficient mice. Neurosci Lett 199:1–4

Gossen M, Bujard H (1992) Tight control of gene expression in mammalian cells by tetracycline-responsive promoters. Proc Natl Acad Sci USA 89:5547–5551

Gozes I, Bachar M, Bardea A, Davidson A, Rubinraut S, Fridkin M, Giladi E (1997) Protection against developmental retardation in apolipoprotein E-deficient mice by a fatty neuropeptide: Implications for early treatment of Alzheimer's disease. J Neurobiol 33:329–342

Green SM, Vonsattel J-PG (1997) Diagnosis of cerebral amyloid angiopathy. Stroke 28:1418–1422

Han S-H, Einstein G, Weisgraber KH, Strittmatter WJ, Saunders AM, Pericak-Vance M, Roses AD, Schmechel DE (1994) Apolipoprotein E is localized to the cytoplasm of human cortical neurons: A light and electron microscopic study. J Neuropathol Exp Neurol 53:535–544

Harris-White ME, Chu T, Sigel JJ, Flanders KC, Frautschy SA. (1998) Effects of the transforming growth factor βs (1–3) on Aβ deposition and injury in hippocampal slice cultures. Neurobiol Aging 19(4S):S267 (#1116)

Hennighausen L, Wall RJ, Tillmann U, Li ML, Furth PA (1995) Conditional gene expression in secretory tissues and skin of transgenic mice using the MMTV-LTR and the tetracycline responsive system. J Cell Biochem 59:463–472

Hyman BT, Gomez-Isla T, Rebeck GW, Briggs M, Chung HY, West HL, Greenberg S, Mui S, Nichols S, Wallace R, Growdon JH (1996) Epidemiological, clinical, and neuropathological study of apolipoprotein E genotype in Alzheimer's disease. Ann NY Acad Sci 802:1–5

Irizarry MC, McNamara M, Fedorchak K, Hsiao K, Hyman BT (1997a) APP_{Sw} transgenic mice develop age-related Aβ deposits and neuropil abnormalities, but no neuronal loss in CA1. J Neuropathol Exp Neurol 56:965–973

Irizarry MC, Soriano F, McNamara M, Page KJ, Schenk D, Games D, Hyman BT (1997b) Aβ deposition is associated with neuropil changes, but not with overt neuronal loss in the human amyloid precursor protein V717F (PDAPP) transgenic mouse. J Neurosci 17:7053–7059

Krugers HJ, Mulder M, Korf J, Havekes L, De Kloet ER, Joëls M (1997) Altered synaptic plasticity in hippocampal CA1 area of apolipoprotein E deficient mice. NeuroReport 8:2505–2510

Lambert MP, Barlow AK, Chromy BA, Edwards C, Freed R, Liosatos M, Morgan TE, Rozovsky I, Trommer B, Viola KL, Wals P, Zhang C, Finch CE, Krafft GA, Klein WL (1998) Diffusible, nonfibrillar ligands derived from $A\beta_{1-42}$ are potent central nervous system neurotoxins. Proc Natl Acad Sci USA 95:6448–6453

Laskowitz DT, Sheng H, Bart RD, Joyner KA, Roses AD, Warner DS (1997) Apolipoprotein E-deficient mice have increased susceptibility to focal cerebral ischemia. J Cereb Blood Flow Metab 17:753–758

Lendon CL, Ashall F, Goate AM (1997) Exploring the etiology of Alzheimer disease using molecular genetics. JAMA 277:825–831

Lomnitski L, Kohen R, Chen Y, Shohami E, Trembovler V, Vogel T, Michaelson DM (1997) Reduced levels of antioxidants in brains of apolipoprotein E-deficient mice following closed head injury. Pharmacol Biochem Behav 56:669–673

Masliah E, Mallory M, Alford M, Ge N, Mucke L (1995a) Abnormal synaptic regeneration in hAPP695 transgenic and ApoE knockout mice. In: Iqbal K, Mortimer J, Winblad B, Wisniewski H (eds.). Research advances in Alzheimer's disease and related disorders. John Wiley, New York, pp. 405–414

Masliah E, Mallory M, Ge N, Alford M, Veinbergs I, Roses AD (1995b) Neurodegeneration in the central nervous system of apoE-deficient mice. Exp Neurol 136:107–122

Masliah E, Sisk A, Mallory M, Mucke L, Schenk D, Games D (1996) Comparison of neurodegenerative pathology in transgenic mice overexpressing V717F β-amyloid precursor protein and Alzheimer's disease. J Neurosci 16:5795–5811

Masliah E, Samuel W, Veinbergs I, Mallory M, Mante M, Saitoh T (1997) Neurodegeneration and cognitive impairment in apoE-deficient mice is ameliorated by infusion of recombinant apoE. Brain Res 751:307–314

Mayford M, Bach ME, Huang YY, Wang L, Hawkins RD Kandel ER (1996) Control of memory formation through regulated expression of a CaMKII transgene. Science 274:1678–1683

Mazur-Kolecka B, Frackowiak J, Haskea T, Le Vine H, III, Krzeslowska J, Golabek A, Wisniewski HM (1998) TGFβ potentiates the intracellular aggregation of Aβ in response to apoE: a possible mechanism to form nuclei of Aβ polymerization. Neurobiol Aging 19(4S):S130 (#548)

McGeer PL, McGeer EG (1995) The inflammatory response system of brain: Implications for therapy of Alzheimer and other neurodegenerative diseases. Brain Res Rev 21:195–218

Mercken L, Brion JP (1995) Phosphorylation of tau protein is not affected in mice lacking apolipoprotein E. NeuroReport 6:2381–2384

Metzger RE, LaDu MJ, Pan JB, Getz GS, Frail DE, Falduto MT (1996) Neurons of the human frontal cortex display apolipoprotein E immunoreactivity: Implications for Alzheimer's disease. J Neuropathol Exp Neurol 55:372–380

Miyata M, Smith JD (1996) Apolipoprotein E allele-specific antioxidant activity and effects on cytotoxicity by oxidative insults and β-amyloid peptides. Nat Genet 14:55–61

Mucke L, Eddleston M (1993) Astrocytes in infectious and immune-mediated diseases of the central nervous system. FASEB J 7:1226–1232

Nathan BP, Bellosta S, Sanan DA, Weisgraber KH, Mahley RW, Pitas RE (1994) Differential effects of apolipoproteins E3 and E4 on neuronal growth in vitro. Science 264:850–852

Nelson RJ (1997) The use of genetic "knockout" mice in behavioral endocrinology research. Hormones Behav 31:188–196

Oitzl MS, Mulder M, Lucassen PJ, Havekes LM, Grootendorst J, De Kloet ER (1997) Severe learning deficits in apolipoprotein E-knockout mice in a water maze task. Brain Res 752:189–196

Ott A, Breteler MMB, van Harskamp F, Stijnen T, Hofman A (1998) Incidence and risk of dementia - the Rotterdam study. Am J Epidemiol 147:574–580

Owen EH, Logue SF, Rasmussen DL, Wehner JM (1997) Assessment of learning by the Morris water task and fear conditioning in inbred mouse strains and F_1 hybrids: implications of genetic background for single gene mutations and quantitative trait loci analysis. Neuroscience 80:1087–1099

Pike CJ, Burdick D, Walencewicz AJ, Glabe CG, Cotman CW (1993) Neurodegeneration induced by β-amyloid peptides *in vitro:* The role of peptide assembly state. J Neurosci 13:1676–1687

Prehn JH, Bindokas VP, Marcuccilli CJ, Krajewski S, Reed JC, Miller RJ (1994) Regulation of neuronal Bcl 2 protein expression and calcium homeostasis by transforming growth factor type β confers wide-ranging protection on rat hippocampal neurons. Proc Natl Acad Sci USA 91:12599–12603

Prehn JHM, Bindokas VP, Jordán J, Galindo MF, Ghadge GD, Roos RP, Boise LH, Thompson CB, Krajewski S, Reed JC, Miller RJ (1996) Protective effect of transforming growth factor-β1 on β-amyloid neurotoxicity in rat hippocampal neurons. Mol Pharmacol 49:319–328

Prior R, Urmoneit B, Prikulis I, Whil G, Fischer J, Frank R, D'Urso D (1998) Alzheimer amyloid angiopathy and transforming growth factor β1. Neurobiol Aging 19(4S):S113 (#478)

Raber J, Wong D, Buttini M, Orth M, Bellosta S, Pitas RE, Mahley RW, Mucke L (1998) Isoform-specific effects of human apolipoprotein E on brain function revealed in *Apoe* knockout mice - increased susceptibility of females. Proc Natl Acad Sci USA 95:10914–10919

Racke MK, Sriram S, Carlino J, Cannella B, Raine CS, McFarlin DE (1993) Long-term treatment of chronic relapsing experimental allergic encephalomyelitis by transforming growth factor-β2. J Neuroimmunol 46:175–184

Roses AD, Saunders AM (1997) Cerebrovascular pathology in Alzheimer's disease. Ann NY Acad Sci 826:200–212

Selkoe D (1997) Alzheimer's disease: genotypes, phenotypes, and treatments. Science 275:630–631

Stone DJ, Rozovsky I, Morgan TE, Anderson CP, Finch CE (1998) Increased synaptic sprouting in response to estrogen via an apolipoprotein E-dependent mechanism: implications for Alzheimer's disease. J Neurosci 18:3180–3185

Strittmatter WJ, Roses AD (1996) Apolipoprotein E and Alzheimer's disease. Annu Rev Neurosci 19:53–77

Sun YL, Wu S, Bu GJ, Onifade MK, Patel SN, LaDu MJ, Fagan AM, Holtzman DM (1998) Glial fibrillary acidic protein-apolipoprotein E (apoE) transgenic mice: astrocyte-specific expression and differing biological effects of astrocyte-secreted apoE3 and apoE4 lipoproteins. J Neurosci 18:3261–3272

Terry RD (1996) The pathogenesis of Alzheimer disease: An alternative to the amyloid hypothesis. J Neuropathol Exp Neurol 55:1023–1025

Turner RS, Suzuki N, Chyung ASC, Younkin SG, Lee VM-Y (1996) Amyloids β_{40} and β_{42} are generated intracellularly in cultured human neurons and their secretion increases with maturation. J Biol Chem 271:8966–8970

van der Wal EA, Gómez-Pinilla F, Cotman CW (1993) Transforming growth factor-β1 is in plaques in Alzheimer and Down pathologies. NeuroReport 4:69–72

Veinbergs I, Jung MW, Young SJ, Van Uden E, Groves PM, Masliah E (1998) Altered long-term potentiation in the hippocampus of apolipoprotein E-deficient mice. Neurosci Lett 249:71–74

Wahl SM, Allen JB, McCartney-Francis N, Morganti-Kossmann MC, Kossmann T, Ellingsworth L, Mai UE, Mergenhagen SE, Orenstein JM (1991) Macrophage- and astrocyte-derived transforming growth factor beta as a mediator of central nervous system dysfunction in acquired immune deficiency syndrome. J Exp Med 173:981–991

Weisgraber KH, Mahley RW (1996) Human apolipoprotein E: The Alzheimer's disease connection. FASEB J 10:1485–1494

Wyss-Coray T, Feng L, Masliah E, Ruppe MD, Lee HS, Toggas SM, Rockenstein EM, Mucke L (1995) Increased central nervous system production of extracellular matrix components and development of hydrocephalus in transgenic mice overexpressing transforming growth factor-β1. Am J Pathol 147:53–67

Wyss-Coray T, Borrow P, Brooker MJ, Mucke L (1997a) Astroglial overproduction of TGF-β1 enhances inflammatory central nervous system disease in transgenic mice. J Neuroimmunol 77:45–50

Wyss-Coray T, Masliah E, Mallory M, McConlogue L, Johnson-Wood K, Lin C, Mucke L (1997b) Amyloidogenic role of cytokine TGF-β1 in transgenic mice and Alzheimer's disease. Nature 389:603–606

Wyss-Coray T, Masliah E, Lin C, Mucke L (1998) Apolipoprotein E: A potential mediator of cytokine-induced cerebral amyloidosis in transgenic mice and humans with Alzheimer's disease. Neurobiol Aging 19(4S):S55 (#229)

Xu PT, Gilbert JR, Qui H, Ervin J, Rothrock-Christian TR, Hulette C, Schmechel DE (1999) Specific regional transcription of apolipoprotein E in human brain neurons. Am J Pathol 154:601–611

Yankner BA, Duffy LK, Kirschner DA (1990) Neurotrophic and neurotoxic effects of amyloid beta Protein: reversal by tachykinin neuropeptides. Science 250:279–282

Zhou Y, Elkins PD, Howell LA, Ryan DH, Harris RBS (1998) Apolipoprotein-E deficiency results in an altered stress responsiveness in addition to an impaired spatial memory in young mice. Brain Res 788:151–159

Cytokine Communication Between Neurons and Glia and the Pathogenesis of Alzheimer's Disease

B. Cordell, L. Yang, A. Naidu, K. Bales, S. Paul, and G. Murphy

Summary

Brain inflammation is a key feature in the development of Alzheimer's disease. Numerous inflammatory markers can be observed in afflicted brain regions by immunohistochemical analysis. Epidemiological studies indicate that exposure to anti-inflammatory drugs can delay the onset of this disease. Microglia are likely to have a pivotal role in this inflammatory process since these cells are potent mediators of cerebral inflammation. Microglia have also been found specifically associated with β-amyloid deposits that progress into mature neuritic plaques, structures that are considered to be intimately connected with the pathology and progression of the disease. We have examined β-amyloid-induced inflammatory responses in a murine microglial cell line, BV-2. We have found that exposure of BV-2 cells to exogenous β-amyloid induces secretion of endogenous β-amyloid. Furthermore, β-amyloid exposure results in pro-inflammatory events, such as the translocation of the transcriptional activator, NF-κB, and a 2- to 5-fold induction of IL-1 and IL-6. Exposure of neurons to β-amyloid peptide has been reported to induce release of macrophage-colony stimulating factor (M-CSF). Because microglia are target cells for M-CSF and because M-CSF is found at increased levels in the central nervous system in Alzheimer's disease, we examined the effects of this cytokine on the inflammatory activation of BV-2 cells. We found that while M-CSF and β-amyloid each have modest effects on BV-2 cytokine production, together these two molecules act synergistically to dramatically generate IL-1, IL-6, and nitric oxide. Identical results have been obtained in rat organotypic hippocampal cultures. Each of these molecules has been shown to injure neurons and to activate astrocytes causing further neuronal injury. These results have been used to build a model for a self-amplifying pathological cascade in Alzheimer's disease involving cytokine communication between neurons, microglia, and astrocytes.

Introduction

Alzheimer's disease is a tragic and prevalent neurodegenerative disorder characterized clinically by the loss of short-term memory and other cogni-

Patterson/Kordon/Christen (Eds.)
Neuro-Immune Interactions in Neurologic and Psychiatric Disorders

tive functions. The disease is defined by the presence of numerous neuropathological lesions, neuritic plaques, which are primarily localized in the hippocampus and cerebral cortex. These anatomical regions are responsible for memory and cognition and the presence of neuritic plaques has provided one explanation for the origin of this disease. The major component of the neuritic plaque is β-amyloid, a 39–43 residue fragment of a larger membrane-spanning glycoprotein referred to as the β-amyloid precursor protein. Considerable attention has been focused on β-amyloid as the possible etiologic agent mediating the observed neurodegeneration. Indeed, much of the available evidence (genetic, biochemical and neuropathological) indicates that β-amyloid is a central and essential molecule in the pathogenesis of Alzheimer's disease (reviewed by Hardy 1997; Selkoe 1997). While the β-amyloid hypothesis has considerable scientific support, it seems that the presence of β-amyloid per se is insufficient to promote the disease process. Hence, this hypothesis needs refinement.

β-Amyloid deposits can be placed into two general categories, diffuse and dense. Neuritic plaques are of the latter category. Closely associated with the neuritic plaque are microglia and astrocytes. Within the plaque are dystrophic neurites, engorged synaptic terminals presumed to be dysfunctional. For diffuse β-amyloid deposits, all of these associated cellular features are absent (Ikeda et al. 1989; Perlmutter et al. 1992). A correlation has been established between the presence of neuritic plaques and the loss of neurons and synapses (Davies et al. 1987; Terry et al. 1991), as well as with the intellectual decline in Alzheimer's disease (Samuel et al. 1994). No such correlation exists for diffuse deposits. The diffuse deposit and neuritic plaque are, however, thought to be related (Ikeda et al. 1989; Griffin et al. 1995). In the initial stages of the disease, only diffuse β-amyloid deposits are observed. As the disease progresses, a subset of these deposits, primarily those in the hippocampus and the parietotemporal cortex, appears to be converted to neuritic plaques. While the molecular signal(s) responsible for the selective conversion of diffuse to dense plaques is not known, the presence of microglia is strongly correlated with this transition, as judged by immunohistochemical analyses. Microglia are a resident population of brain macrophage-like cells that are potent mediators of cerebral inflammation in a variety of disease states. Diverse pathological insults result in activation of microglia. Once activated, a complex set of biochemical and morphological alterations occurs, resulting in the generation of acute phase reactants, proteins associated with the classical complement cascade, inflammatory and chemotactic cytokines, and the release of reactive oxygen and nitrogen species (reviewed by Mrak et al. 1995; Cotman et al. 1996). Interleukin-1 (IL-1), interleukin-6 (IL-6), and tumor necrosis factor-α (TNF-α), are examples of the pro-inflammatory mediators produced by activated microglia within the neuritic plaque. Activated microglia are an early and constant cell component of the neuritic plaque (Sheng et al. 1997; Griffin et al. 1995). In contrast, in anatomical brain areas that are not involved in the behavioral manifestations of Alzheimer's disease, such as the cerebellum, diffuse β-amyloid deposits are present but activated microglia are absent (Perlmutter et al. 1992; Sheng et al. 1997). Because neuritic plaques typically do not form in the cerebellum (Ikeda et al. 1989, this observation indirectly supports the critical role for microglia in the conversion of diffuse deposits to the pathogenic neuritic plaque.

Astrocyte involvement is also unique to the neuritic plaque (reviewed by Mrak et al. 1995; Cotman et al. 1996). This cell type has been observed to enter the histological progression after the initial infiltration of microglia, and astrocytes may be recruited to the plaque vicinity by signals from activated microglia and possibly from injured neurons. IL-1, released by activated microglia, can exert paracrine effects on astrocytes including activation and the resultant up-regulation of a variety of molecules such as IL-6, TNF-α, α1-antichymotrypsin and α-2-macroglobulin, both acute phase reactant proteins (reviewed by Cotman et al. 1996; Merrill and Benveniste 1996), and apolipoprotein E, a protein that contributes to Alzheimer's disease risk (Strittmatter and Roses 1996). In addition to astrocyte activation, there is also hypertrophy. This astrocytosis is seen as a halo of cells surrounding the neuritic plaque. Similar to the microglial profile, there is a much lower frequency of activated astrocytes affiliated with diffuse deposits of β-amyloid, especially in the cerebellum.

Epidemiological studies, both retrospective (reviewed by McGeer et al. 1996) and prospective (Stewart et al. 1997), evaluating the use of non-steroidal anti-inflammatory drugs (NSAIDs) provide further evidence for the importance of the inflammatory events in Alzheimer's disease. Exposure to NSAIDs has been found to exhibit a protective effect against the appearance of the disease and a beneficial effect on Alzheimer's patients with respect to rate of cognitive decline (Rogers et al. 1993). Based on these and other data, an inflammatory hypothesis of Alzheimer's disease has been put forward which posits that chronic cerebral inflammation results in injury to neurons, contributing over time to cognitive decline. Neuronal injury is hypothesized to result from the effects of inflammatory effectors produced by glia, such as cytokines and reactive oxygen and nitrogen species.

In vitro Activation of Microglia by β-Amyloid

From these and other data, extensive evidence exists indicating an inflammatory reaction in Alzheimer's disease brain. However, a number of issues remain unresolved: What is the origin of this inflammatory process? What is the molecular signal(s) stimulating the initiation of the inflammatory response? How is the inflammatory process related to the select neuronal degeneration seen in the disease? Which aspects of the inflammatory response are fundamental to disease progression and which are epiphenomena? While complete answers to these questions are not yet available, pieces are being assembled from in vitro experimentation. From the collective research in the field, β-amyloid has emerged as one likely contributing factor to the inflammatory component of this disease.

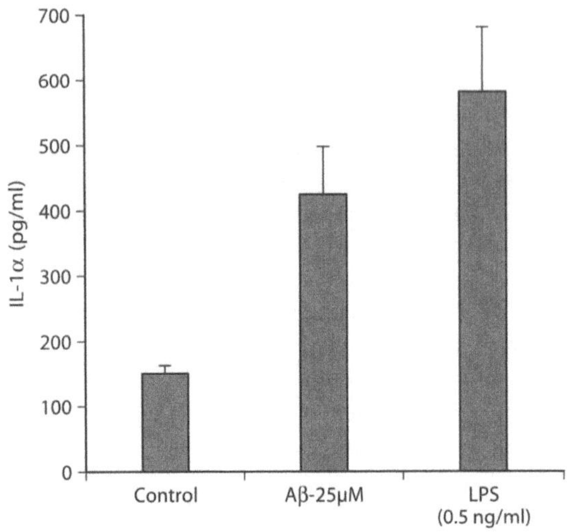

Fig. 1. Induction of IL-1α in BV-2 microglia by β-amyloid (Aβ) and by lipopolysaccharide (LPS)

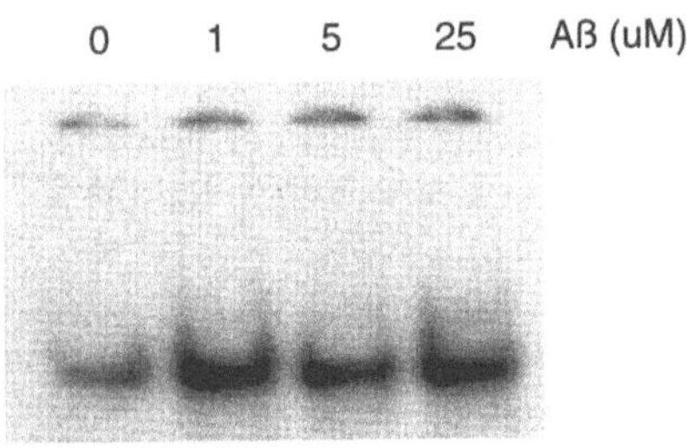

Fig. 2. Dose-dependent activation of NF-κB in nuclear extracts prepared from BV-2 cells treated with β-amyloid (Aβ)

β-Amyloid has been demonstrated to be capable of eliciting inflammatory responses in cultured cells. For example, when applied to primary rat microglia, β-amyloid was found to trigger the production of IL-1 (Araujo and Cotman 1992). Experiments conducted in our laboratories using the immortalized murine microglial cell line, BV-2 (Bocchini et al. 1992), yielded similar results. Figure 1 shows the several-fold induction of IL-1α typically observed for microglia treated with 10–25 μM β-amyloid. We found a comparable induction of IL-6 in identical experiments. Consistent with cytokine production as a measure of BV-2 microglial activation, we also observed the nuclear translocation of the transcriptional activator, NF-κB, following β-amyloid treatment (Fig. 2). NF-κB has a well-documented role in response to a vast array of stimuli, which is mediated at the transcriptional level (reviewed in Stancovski and Baltimore 1997). The genes regulated by the NF-κB transcription factor are diverse and include those involved in immune function, inflammatory response, cell adhesion, and growth control (reviewed in Baldwin 1996).

Further support for β-amyloid activation of microglia is the study reported by Meda et al. (1995). This group demonstrated the release of TNF-α

and reactive nitrogen species when β-amyloid was applied in combination with γ-interferon to the N9 immortalized murine microglial cell line (Meda et al. 1995; Ii et al. 1996). Another relevant feature of β-amyloid has been noted, i.e., the peptide is chemotactic for microglia (Davis et al. 1992). The combined chemotactic and cylokine-inducing properties of β-amyloid for microglia have clear implications with respect of Alzheimer's histopathology. Lacking, however, is an explanation for the select infiltration of microglia into diffuse β-amyloid deposits in the hippocampus and cortex as opposed to those in the cerebellum. Other signals, in addition to β-amyloid, must be participating in vivo.

Each of the microglial cytokines induced by β-amyloid, namely IL-1, IL-6 and TNF-α, have well-documented proliferation and/or activation effects for cultured astrocytes (reviewed in Merrill and Benveniste 1996). Therefore, it is logical to postulate that these same cytokines have an active role in the astrocytosis observed at the site of the neuritic plaque. In addition to these possible indirect effects of β-amyloid on astrocytes, β-amyloid also has direct pro-inflammatory effects. The peptide has been demonstrated to potentiate IL-6 and IL-8 release by IL-1β from a human astrocytoma cell line (Gitter et al. 1995). In addition, β-amyloid has been demonstrated to induce nitric oxide synthase (iNOS) expression and nitric oxide (NO) generation by primary rat astrocytes in culture (Hu et al. 1998).

The β-amyloid-induced increases in microglial production of cytokines have, however, been of limited magnitude, on the order of only 2- to 5-fold greater than control levels. Consequently, it is difficult to reconcile this weak in vitro microglial response to β-amyloid with the hypothesis that β-amyloid activation of microglia is important in Alzheimer's disease pathophysiology. One reason for the limited response of cultured microglia to β-amyloid may be that important co-stimulatory agents present in Alzheimer's disease brain have not been taken into consideration in prior reports. The extracellular environment surrounding neuritic plaques in Alzheimer's disease brain is rich in a variety of pro-inflammatory agents, including cytokines (reviewed in Cotman et al. 1996), which are likely to augment the effects of β-amyloid on microglia. While it has been shown that interferon-γ can augment the effects of β-amyloid on microglia (Meda et al. 1995), the synergistic increase in β-amyloid-induced microglial activity in cultures co-treated with this stimulus is not likely to have direct relevance to Alzheimer's disease due to the lack of involvement of peripheral B and T cells in this disease process. The ability of a cytokine to act synergistically with β-amyloid, however, did prompt us to consider the possibility of other co-stimulators of β-amyloid-induced microglia activation. We reasoned that macrophage colony stimulating factor (M-CSF; also termed colony stimulating factor 1; CSF-1) might be a relevant partner in the β-amyloid activation of microglia.

M-CSF: β-Amyloid Co-Stimulatory Effects on Microglia

M-CSF is an important regulator of mononuclear phagocyte development and function throughout the body. In the brain, M-CSF is expressed by neurons, astrocytes, and endothelial cells (Yan et al. 1997; Hao et al. 1990; Thery et al. 1992) where it induces proliferation, migration, and activation of microglia (see Merrill and Benveniste 1996). M-CSF treatment of microglia also induces increased expression of macrophage scavenger receptors (Yan et al. 1997) that mediate microglial interactions with β-amyloid (Paresce et al. 1996; Khoury et al. 1996). β-Amyloid binds to neuronal receptors for advanced glycation end-products to increase neuronal M-CSF expression (Yan et al. 1997), which causes further microglial activation. In Alzheimer's disease brain there is increased immunoreactivity for the M-CSF receptor, c-fms, on microglia (Akiyama et al. 1994), Neurons in Alzheimer's disease show labeling with M-CSF antibodies, and M-CSF levels in Alzheimer's disease cerebrospinal fluid are 5-fold greater than in controls (Yan et al. 1997; Singh and Guthikonda 1997). Thus, M-CSF represents a potent microglial activator relevant to Alzheimer's disease pathophysiology. We hypothesized that in Alzheimer's disease M-CSF activates microglia to augment β-amyloid-induced production of inflammatory cytokines and reactive nitrogen species, which in turn promote additional inflammation, as well as neuronal injury. We tested this hypothesis by examining the effects of combined M-CSF and β-amyloid treatment on the production of IL-1, IL-6 and NO by the BV-2 immortalized murine microglial cell line. BV-2 cells express receptors for advanced glycation end products that bind β-amyloid and induce signal transduction, and BV-2 cells treated with M-CSF show chemotaxis and other indications of activation (Yan et al. 1997). Hence, the BV-2 cell line bears the appropriate molecular machinery to respond to M-CSF.

We treated BV-2 microglial cells with 11 μM synthetic β-amyloid 1-40. This resulted in increased IL-1*α* production to approximately 3-fold over control levels. However, when BV-2 cells were simultaneously treated with M-CSF (25 or 50 ng/ml) and 11 μM β-amyloid 1-40, there was a large increase in IL-1*α* production; approximately 70-fold above control levels (Fig. 3). M-CSF alone had little effect on IL-1*α* production by the BV-2 cells. β-Amyloid 40-1, a reverse sequence control peptide that was prepared in the same manner as β-amyloid 1-40, had little effect either alone or in combination with M-CSF. Neither M-CSF alone nor in combination with β-amyloid resulted in proliferation of BV-2 cells at the doses, cell density, and treatment duration used in the study. Additional evidence for the augmenting effect of M-CSF on β-amyloid-induced IL-1 production by BV-2 cells was that the synergistic effect could be inhibited by a monoclonal antibody to the mouse M-CSF receptor, c-fms, but not by a subclass-matched control antibody.

Simultaneous treatment of BV-2 cells with M-CSF and β-amyloid also induced a very large increase in IL-6 production. Treatment of BV-2 cells with M-CSF alone or β-amyloid alone resulted in modest increases in mouse IL-6 in conditioned media. However, the combination of M-CSF and β-amyloid

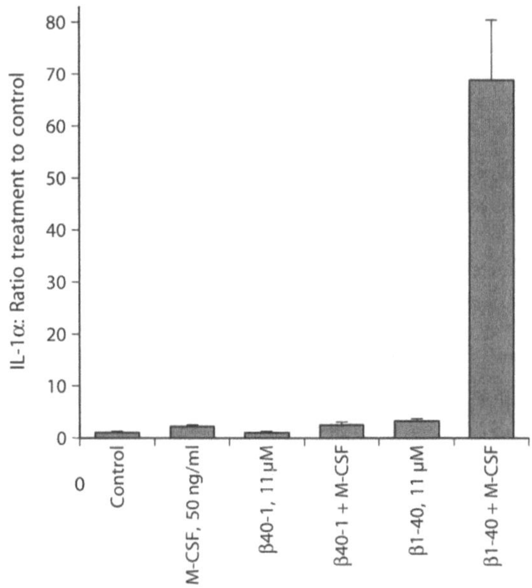

Fig. 3. Synergistic induction of IL-1*a* in BV-2 microglia following co-treatment with *β*-amyloid (A*β*) and M-CSF

resulted in an increase in IL-6 production by BV-2 cells which was over 200-fold greater than control values.

M-CSF also augmented *β*-amyloid effects on NO production monitored as nitrite. Treatment of BV-2 cells with *β*-amyloid 1-40, or M-CSF alone had little effect on nitrite in conditioned medium. In contrast, simultaneous treatment of BV-2 cells with *β*-amyloid and M-CSF resulted in nitrite levels in conditioned medium that were 30-fold greater than control values. Again, the control peptide *β*-amyloid 40-1, either alone or in combination with M-CSF, had little effect on nitrite levels in conditioned medium. The augmenting effect of combined *β*-amyloid and M-CSF treatment on microglial NO production was also found to increase in iNOS mRNA, in contrast to each agent alone.

To further test for the specificity of M-CSF inaugmenting *β*-amyloid activity on microglia, other related cytokines, GM-CSF and IL-3, were tested, each alone or in combination with *β*-amyloid. Both GM-CSF and IL-3 can induce microglial activation (Giulian and Ingeman 1988). Whereas GM-CSF and IL-3 share a common receptor subunit (β_c), the M-CSF receptor, c-fms, is distinct (Hamilton 1997). Only M-CSF possessed augmenting effects of *β*-amyloid on the BV-2 microglia. Differences in receptor function and signal transduction between M-CSF and other hematopoietic cytokines may account for the unique effects of M-CSF on *β*-amyloid-treated microglia.

Our results indicate that *β*-amyloid has a pro-inflammatory action on microglia and that these effects are greatly pronounced when in the presence of

M-CSF, a cytokine shown to be increased in Alzheimer's disease brain and cerebrospinal fluid. Increases in IL-1 and IL-6 production in Alzheimer's disease may be due to the paired exposure of microglia to β-amyloid and M-CSF, resulting in large increases in these cytokines. IL-1 is expressed early in Alzheimer's disease primarily by microglia (Griffin et al. 1995; Sheng et al. 1997) and, through autocrine and paracrine mechanisms, could further augment microglial-mediated inflammation and local neuronal injury in Alzheimer's disease. IL-6, another microglial cytokine, is also increased in Alzheimer's disease brain (Bauer et al. 1991) and in the serum of patients (Singh and Guthikonda 1997). Increased IL-6 expression may induce inflammatory changes that injure neurons. There is evidence that NO, an important inflammatory effector produced by rodent and human microglia, is present at increased levels in Alzheimer's disease brain, resulting in nitration of proteins and other abnormal cellular changes (Smith et al. 1997).

We extended these observations to an experimental system that more closely recapitulates the brain, organotypic hippocampal slice cultures. When 21-day-old rat hippocampal cultures were treated with only β-amyloid, a modest increase in mRNAs encoding IL-1, IL-6, and iNOS was seen. Correspondingly, changes in IL-1 and NO generation were small. β-Amyloid and M-CSF co-treatment, however, resulted in an increase in expression of IL-1, IL-6, and NO that was larger than that observed with either agent alone. We might predict neuronal injury to accompany the microglial activation prompted by the β-amyloid: M-CSF treatment, since exogenous application of β-amyloid-activated macrophage to rat hippocampal slices has been reported to elicit neurotoxicity in the cultured tissue (London et al. 1996).

Model of Glial and Neuronal Pathological Interaction

These results and the results of prior studies indicate that, in Alzheimer's disease, there is a self-amplifying pathophysiologic cascade involving microglia, astrocytes, and neurons and the key Alzheimer's disease cytokines M-CSF, IL-1, and IL-6 (Fig. 4; Murphy et al. 1998). M-CSF levels are increased in the cerebrospinal fluid of Alzheimer's disease patients and M-CSF antibodies label neurons in Alzheimer's disease brain (Yan et al. 1997). Further, expression of the M-CSF receptor c-fms is increased on microglia in Alzheimer's disease brain (Akiyama et at. 1994), which may sensitize these cells to M-CSF effects. Astrocytes, an important source of M-CSF in the brain, can be induced to secrete M-CSF by IL-1 (Thery et al. 1992). We hypothesize that, in Alzheimer's disease, β-amyloid induces microglia to secrete small amounts of IL-1, as our results and the results of others indicate. IL-1 then induces astrocytes to express M-CSF, which augments β-amyloid-induced expression of IL-1 by microglia, resulting in further M-CSF expression by astrocytes. In addition, microglial IL-1 self-activates microglia via autocrine and paracrine effects. Neurons themselves may also secrete M-CSF in response to β-amyloid (Yan et al. 1997), which may further activate microglia.

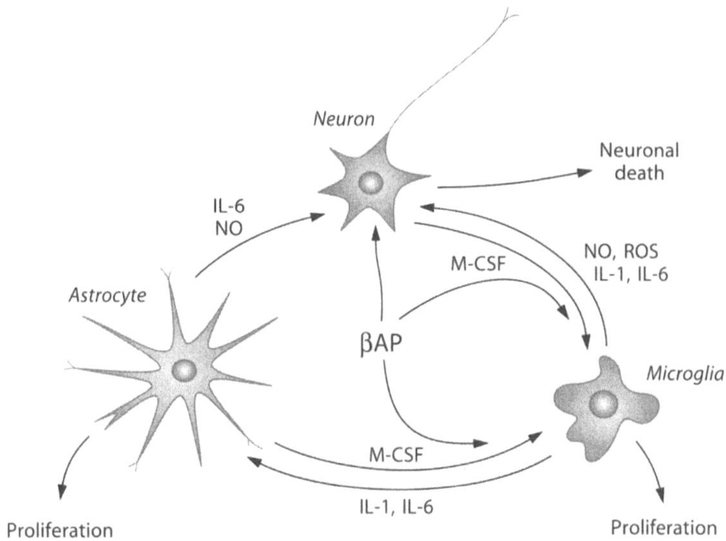

Fig. 4. Model for a self-amplifying pathologic cascade in Alzheimer's disease involving the inflammatory interactions between microglia, astrocytes, and neurons

IL-6 promotes astrogliosis and activates microglia (see Merrill and Benveniste 1996). Increased IL-6 found in Alzheimer's disease brain could come from either microglia or astrocytes, or both. Our results suggest that M-CSF and β-amyloid would induce microglial IL-1 and IL-6 production in Alzheimer's disease. IL-1 causes astrocytes to express IL-6, so microglial IL-1 induced by M-CSF and β-amyloid would promote astroglial IL-6 expression. Through pro-inflammatory effects, IL-6 is thought to contribute to neurodegeneration in Alzheimer's disease.

Although β-amyloid alone may increase microglial NO production, β-amyloid-induced microglial NO production is dramatically elevated, in the presence of M-CSF. Contrary to prior findings, recent evidence indicates that human microglia can produce NO (Ding et al. 1997), so results obtained with murine cells are likely to closely model the human system. Microglial NO, either directly or through its highly toxic derivative peroxynitrite, would injure neurons in Alzheimer's disease (Ii et al. 1996). NO may also induce additional IL-1 expression (see Merrill and Benveniste 1996), which in turn would promote astroglial M-CSF expression, ultimately resulting in further β-amyloid-induced NO production. Astrocytes, too, produce NO and IL-1 can induce astrocyte iNOS (Lee et al. 1993). Thus, in Alzheimer's disease, microglial IL-1 induced by β-amyloid and M-CSF would augment NO neurotoxicity by activating astrocyte iNOS. Microglia are also likely to generate neurotoxic reactive oxygen species in response to β-amyloid (Klegeris et al. 1994).

Finally, another aspect of the proposed model is that β-amyloid can stimulate its own secretion from microglia. We have found that exposure of BV-2 cells to exogenous synthetic β-amyloid results in increased release of the endogenous β-amyloid from the cell (Bitting et al. 1996). This finding suggests that microglial-derived β-amyloid could further contribute to the amplification cascade, either through paracrine effects on neurons to induce the release of M-CSF or through autocrine effects, in conjunction with M-CSF, to stimulate IL-1, IL-6, and NO generation.

Hypotheses Revisited

Two major hypotheses of Alzheimer's disease pathogenesis have been put forward, the amyloid hypothesis and the inflammatory hypothesis. Both concepts are supported by scientific data. Current data suggest, however, that these hypotheses may be directly linked in that β-amyloid has emerged as a key contributor to the inflammatory response seen in the Alzheimer's brain. Therefore, it seems reasonable that these two concepts can be unified into one hypothesis by claiming a pivotal role for β-amyloid in the chronic, neurotoxic inflammatory processes active in this disease.

References

Akiyama H, Nishimura T, Kondo, H, Ikeda K, Hayashi Y, McGeer PL (1994) Expression of the receptor for macrophage colony stimulating factor by brain microglia and its upregulation in brains of patients with Alzheimer's disease. Brain Res 639:171–174

Araujo DM, Cotman CW (1992) β-Amyloid stimulates glial cells in vitro to produce growth factors that accumulate in senile plaques in Alzheimer's disease. Brain Res 569:141–145

Baldwin AS (1996) The NF-κB and IκB proteins: new discoveries and insights. Annu Rev Immunol 14:649–681

Bauer J, Strauss S, Schreiter-Gasser U, Ganer U, Schlegel P, Witt I, Yolk B, Berger M (1991) Interleukin-6 and alpha-2-macroglobulin indicate an acute-phase state in Alzheimer's disease cortices. FEBS Lett 285:111–114

Bitting L, Naidu A, Cordell B, Murphy GM (1996) β-Amyloid peptide secretion by a microglial cell line is induced by β-amyloid (25-35) and lipopolysaccharide. J Biol Chem 271:16084–16089

Bocchini V, Mazzolla R, Barluzzi R, Blasi E, Sick P, Kettenmann H (1992) An immortalized cell line expressed properties of activated microglial cells. J Neurosci Res 31:616–621

Cotman CW, Tenner AJ, Cummings BJ (1996) β-Amyloid converts an acute phase injury response to chronic injury response. Neurobiol Aging 17:723–731

Davies CA, Mann DMA, Sumpter PQ, Yates PO (1987) A quantitative morphometric analysis of the neuronal and synaptic content of the frontal and temporal cortex in patients with Alzheimer's disease. J Neurol Sci 78:151–164

Davis JB, McMurray HF, Schubert D (1992) The amyloid beta-protein of Alzheimer's disease is chemotactic for mononuclear phagocytes. Biochem Biophys Res Commun 189:1096–1100

Ding M, St. Pierre BA, Parkinson JF, Medberry P, Wong JL, Rogers NE, Ignarro LJ, Merrill JE (1997) Inducible nitric oxide synthase and nitric oxide production in human fetal astrocytes and microglia: a kinetic study. J Biol Chem 272:11327–11335

Gitter BD, Cox LM, Rydell RE, May PC (1995) Amyloid β peptide potentiates cytokine secretion by interleukin 1β-activated human astrocytoma cells. Proc Natl Acad Sci USA 92:10738–10741

Giulian D, Ingeman JE (1988) Colony stimulating factors as promoters of ameboid microglia. J Neurosci 8:4707–4717

Griffin WST, Sheng JG, Roberts GW, Mrak RE (1995) Interleukin-1 expression in different plaque types in Alzheimer's disease: significance in plaque evolution. J Neuropathol Exp Neurol 54:276–281
Hamilton J (1997) CSF-1 singal transduction: what is of functional significance? Immunol Today 18:313–317
Hao C, Guilbert LJ, Federfoo S (1990) Production of colony-stimulating factor-1 (CSF-1) by mouse astroglia in vitro. J Neurosci Res 27:314–323
Hardy J (1997) Amyloid, the presenilins and Alzheimer's disease. Trends Neurosci 20:154–159
Hu J, Akama KT, Krafft GA, Chromy BA, Van Eldik LJ (1998) Amyloid β peptide activates cultured astrocytes: morphological alterations, cytokine induction and nitric oxide release. Brain Res 785:195–206
Ii M, Sunamoto M, Ohnishi K, Ichimori Y (1996) β-Amyloid protein-dependent nitric oxide production from microglial cells and neurotoxicity. Brain Res 720:93–100
Ikeda SI, Allsop D, Glenner GG (1989) Morphology and distribution of plaque and related deposits in the brain of Alzheimer's disease and control cases. Lab Invest 60:113–122
Khoury JE, Hickman SE, Thomas CA, Cao L, Silverstein S, Loike JD (1996) Scavenger receptor-mediated adhesion of microglia to β-amyloid fibrils. Nature 382:716–719
Klegeris A, Walker DG, McGeer PL (1994) Activation of macrophages by Alzheimer β-amyloid peptide. Biochem Biophys Res Comm 199:984–991
Lee SC, Liu W, Roth P, Dickson D, Berman JW, Brosnan CF (1993) Induction of nitric oxide synthase in human astrocytes by interleukin-1β and interferon-γ. J. Immunol 150:594–604
London JA, Biegel D, Patcher JS (1996) Neurocytopathic effects of β-amyloid-stimulated monocytes: a potential mechanism for central nervous system damage in Alzheimer's disease. Proc Natl Acad Sci USA 93:4147–4152
McGeer PL, Schulzer M, McGeer EG (1996) Arthritis and anti-inflammatory agents as possible protective factors for Alzheimer's disease: a review of 17 epidemiologic studies. Neurol 47:425–432
Meda L, Cassatella MA, Szendrel GI, Otvos L, Baron P, Villalba M, Ferrari D, Rossi F (1995) Activation of microglial cells by β-amyloid protein and interferon-γ. Nature 374:647–650
Merrill JE, Benveniste EN (1996) Cytokines in inflammatory brain lesions: helpful and harmful. Trends Neurosci 12:331–338
Mrak RE, Sheng JG, Griffin WST (1995) Glial cytokines in Alzheimer's disease: review and pathologic implications. Human Pathol 26:816–823
Murphy GM, Ynag L, Cordell B (1998) Macrophage colony-stimulating factor augments β-amyloid-induced interleukin-1, interleukin-6, and nitric oxide production by microglia cells. J Biol Chem 273:20967–20971
Paresce DM, Ghosh RN, Maxfield FR (1996) Microglial cells internalize aggregates of the Alzheimer's disease amyloid β-protein via a scavenger receptor. Neuron 17:553–565
Perlmutter LS, Scott SA, Barron E, Chui HC (1992) MHC class II-positive microglia in human brain: association with Alzheimer's lesions. J Neurosci Res 33:549–558
Rogers J, Kirby LC, Hempelman SR, Berry DL, McGeer PL, Kaszniak AW, Zalinski J, Cofield M, Mansukhani L, Willson P, Kogan F (1993) Clinical trials of indomethacin in Alzheimer's disease. Neurology 43:1609–1611
Samuel W, Terry RD, De Teresa R (1994) Clinical correlates of cortical and nucleus basalis pathology in Alzheimer's disease. Arch Neurol 51:772–778
Selkoe D (1997) Alzheimer's disease: genotypes, phenotypes, and treatments. Science 275:630–631
Sheng JG, Mrak RE, Griffin WST (1997) Neuritic plaque evolution in Alzheimer's disease is accompanied by transition of activated microglia from primed to phagocytic forms. Acta Neuropathol 94:1–5
Singh VK, Guthikonda P (1997) Circulating cytokines in Alzheimer's disease. J Psychiat Res 6:657–660
Smith MA, Richey-Harris PL, Sayre LM, Beckman JS, Perry G (1997) Widespread peroxynitrite mediated damage in Alzheimer's disease. J Neurosci 17:2653–2657
Stancovski I, Baltimore D (1997) NF-κB activation: the IκB kinase revealed? Cell 91:299–302
Stewart WF, Kawas C, Corrada M, Metter EJ (1997) Risk of Alzheimer's disease and duration of NSAID use. Neurol 48:626–632

Strittmatter WJ, Roses AD (1996) Apolipoprotein E and Alzheimer's disease. Annu Rev Neurosci 19:53–77

Terry RD, Masliah E, Salmon DP, Butters N, De Teresa R, Hill R, Hansen LA, Katzman R (1991) Physical basis of cognitive alterations in Alzheimer's disease: synapse loss is the major correlate of cognitive impairment. Ann Neurol 30:572–580

Thery C, Stanley ER, Mallat MJ (1992) Interleukin-1 and tumor necrosis factor-alpha stimulate the production of colony stimulating factor 1 by murine astrocytes. J Neurochem 59:1183–1186

Yan SD, Zhu H, Fu J, Yan SF, Roher A, Tourtellotte WW, Rajavashisth T, Chen X, Godman GC, Stern D, Schmidt AM (1997) Amyloid-β peptide-receptor for advanced glycation endproduct interaction elicits neuronal expression of macrophage-colony stimulating factor: a proinflammatory pathway in Alzheimer's disease. Proc Natl Acad Sci USA 94:5296–5301

Yan SD, Chen X, Fu J, Chen M, Zhu H, Roher A, Slattery T, Zhao L, Nagashima M, Morser J, Migheli A, Nawroth P, Stern D, Schmidt AM (1996) RAGE and β-amyloid peptide neurotoxicity in Alzheimer's disease. Nature 382:685–691

Glial Hyperactivity During Aging as a Neuroinflammatory Process

C.E. Finch, T.E. Morgan, Z. Xie, D. Stone, A.-S. Lanzrein, and I. Rozovsky

Summary

We survey evidence from our laboratory and others which shows many indicators of inflammmatory processes during normal aging processes in the brain, particularly increased expression of complement genes and glial hyperactivity. In rodents, activation of microglia and astrocytes occurs by middleage, without any indication of neuropatholoy. However, many of these same markers are associated with the senile plaques of Alzheimer's disease (AD). We hypothesize that endogenous aging processes that activate glia are a factor in the age-related increase of AD.

Introduction

Inflammatory processes during AD are well recognized from many studies. By immunocytochemistry and in situ hybridization, senile plaques display more than 30 inflammatory markers, including complement (C)- system factors, cytokines, and acute phase proteins as described in Table 1 of O'Banion and Finch (1996: special issue of the Neurobiology of Aging) and in the chapter by Joseph Rogers and Young Shen (this volume). Senile plaques also contain highly aggregated amyloid β-peptide (Aβ). Activated astrocytes and microglia are also found in and around senile plaques. Although the specific cellular source of each of the inflammatory proteins described in Table 1 is not known, many of these proteins are encoded by mRNAs, as detected by in situ hybridization in resident brain neurons, astrocytes and microglia (references in Table 1). The idea that resident brain neurons make complement C1q , the initiating molecule of the classical cascade, is still very strange to many immunologists, but this finding from our laboratory with rat and human brain tissue and cultured neurons (Rozovsky et al. 1994; Lampert-Etchells et al. 1993) has been verified (Shen et al. 1997). Moreover, the lesioned brain de novo synthesizes bioactive C1q (Goldsmith et al. 1997).

The important complement receptors are found in neurons, e.g., C5a receptor mRNA (Farkas et al. 1998). The C5 story may be encrypted in the studies of many who inadvertently use common mouse strains that have a frame-shift mutation in the C5 gene, which increases their sensitivity to

Patterson/Kordon/Christen (Eds.)
Neuro-Immune Interactions in Neurologic and Psychiatric Disorders

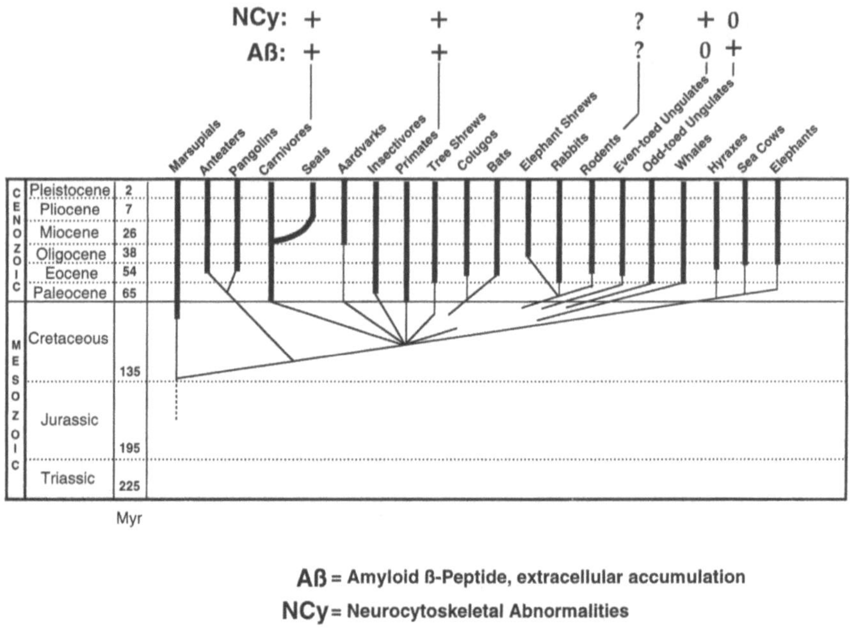

Fig. 1. Distribution of AD-like neuropathology in mammalian orders. Adapted from Finch (1994) and Finch and Sapolsky (in preparation). When information is lacking for an order, no symbol is given. The solid lines approximate the depth of the fossil evidence. The distribution of species that show age-related increases in AD-like neuropathology, as either extracellular Aβ aggregates in senile plaques or cerebrovasculature, or neurocytoskeletal abnormalities, which in human brains are referred to as neurofibrillary tangles (NFT)

neurotoxins and which may play a role in differences between the sensitivity of transgenic hosts to amyloid and other inserts (Pasinetti et al. 1996).

What is *not* found in AD brains is equally interesting: B cells, T cells, or deposits of immunoglobulins (Finch and Marchalonis 1996; Rogers et al. 1988; Rogers and Shen, this volume). Thus, the processes leading to the senile plaque differ fundamentally from the autoimmune-generated plaques of multiple sclerosis. We argued elsewhere that the inflammatory features of AD represent a primitive host defense reaction such as is found in existing organisms that lack somatic cell immunoglobulin diversification mechanisms (Finch and Marchalonis 1996).

A major mystery in mechanisms of AD is the association with aging. An important insight was given by two recent studies that examined the effect of aging on neural responses to injections of aggregated Aβ. In several species of monkeys, the glial reactions and neuronal damage were much greater in older animals (Geula et al. 1998; McKee et al. 1998). We therefore propose the hypothesis that age-related increases in glial reactivity to the amyloid β-peptide (Aβ) are a primary cause of AD. All mammals that live more than 10 years accumulate extracellular aggregates of Aβ and develop neurocytoskeletal abnormalities (Fig. 1). For example, the mouse lemur, a prosimian, shows

Table 1. Neuroinflammatory changes in Alzheimer's and Aging

Increased activity	Senile plaques[a]	Aging rodent[b]	Aging human
Astrocytes	X	X (GFAP)	X (GFAP)[c]
Microglia	X	X (MHC II, CR3)	X
Neuritic Abnormalities	X	X (no NFT)	X
$A\beta$	X		X
APP	X		
α_1-Antichymotrypsin	X		
α_2-Macroglobulin	X		
apoE	X	X	
apoJ (clusterin)	X	X	corpora amylacea[d]
CRP	X		
Heme oxygenase-1	X	X	
Complement factors			
C1q	X	X	CSF[e], corp amylacea[d]
C3	X		
C9	X		corpora amylacea
Cytokines			
IL-1	X		
Il-6	X		
TGF-β1	X	X	
TNFγ	X		

[a] Finch an Marchalonis (1996); [b] Morgan et al. (1999); [c] Nichols et al. (1995); [d] Loeffler et al. (1997); [e] Singhrao et al (1995).

neurocytoskeletal abnormalities by two years of age and $A\beta$ deposits by six to eight years (Bons et al. 1995; Finch and Sapolsky 1999). All other primates examined develop similar AD-like changes by the last third of their maximum life spans (Cork et al. 1990; Gearing et al. 1994; Finch and Sapolsky 1999). These AD-like changes thus qualify as a canonical change of aging in placental mammals that live at least five years (Finch 1990). Laboratory rodents, which rarely live more than 4 years, do not accumulate the $A\beta$ peptide or senile plaque-like material unless carrying human transgenes (Jucker et al. 1992). In view of the ability of the $A\beta$ peptide to activate C1q (Rogers et al. 1992; Velazquez et al. 1997), we propose that the evolution of AD includes an age-related component that may be more primitive evolutionarily than the inflammatory component of aging mediated by amyloid deposits of aging.

Our data provide ample evidence of age changes in glial functions which support this hypothesis. As shown in Table 1, some of the same inflammatory components of senile plaques also increase during aging in rodents, which do not develop senile plaques (Perry et al. 1993; Scott and Mandybur 1996) or in aging primates, which do (Bons et al. 1995; Cork et al. 1990). Brain glia show progressive activation during aging as measured by glial fibrillary acidic protein (GFAP), apolipoprotein E (apoE), apoJ (clusterin), hemeoxygenase-1 (HO-1), complement 3 receptor (CR3), C1qB, major histocompatibility complex II (MHC II), and transforming growth factor (TGF)-β1 (Morgan et al. 1999; Pasinetti et al. 1999). Other examples of increased complement expression or activation during aging include the presence of corpora amylaceae in the brains of cognitvely normal elderly (Singhrao et al. 1995) and increased levels of C3a in the cerebrospinal fluid (Loeffler et al. 1997). Because subsets of these markers occur in peripheral inflammation, we hypothesize that brain aging has inflammatory features. We note that apoJ, which is secreted by astrocytes, promotes the aggregation of Aβ into toxic diffusible forms (ADDLs, amyloid-derived diffusible ligands; Oda et al. 1995; Lambert et al. 1998).

Glial activation arises during middle age in humans and rodents, i.e., in the age group before tumors, kidney lesions, and occlusive vascular conditions and other later, age-specific pathology become prevalent in rodents (Finch 1990). Unlike that in AD, glial activation of normal middleage is not associated with overt neurodegeneration. The glial activation in rodents during normal aging can be dissociated from the vascular and amyloid-related pathology that is very common during aging in humans, because rodents do not accumulate the Aβ peptide.

Experimental Studies on Glial Aging: *in vitro* Models

Moreover, astrocytes and microglia cultured from aging rodent brains show many of these phenotypes (Rozovsky et al. 1998). Our laboratory has developed in vitro cell systems to analyze the complex mechanisms that may be at work in the brain. F344 male rats at various adult ages were used as donors for establishing cultures. Rats aged 12 months can be considered middle-aged, because of the decline of reproduction in both sexes, but in the absence of pathological changes in the brain or other organs. In this genotype, gross organ pathology and accelerating spontaneous mortality are found by 24 months, which defines the phase of senescence. Mixed cultures were originated from microdissected cortex of three to 30-months-old-rats. The identical number of cells were plated for each age group into poly-lysine coated chambers at a density of 2×10^4 viable cells cm^{-2} in DMEM/F12 culture medium supplemented with 10% fetal bovine serum. Media were renewed every other day. After the first 48 hours, cultures were rinsed twice to remove debris and non-adherent cells. Glial cell types were identified by immunocytochemistry. Glial proliferation increased with donor age. After seven days in vitro, [^{3}H]-thymidine incorporation by microglia progressively increased with donor age from aged brains, up to 400% greater at 24 months versus three months. Astrocytes also showed progressive but smaller increases of proliferation with donor age. Moreover, [^{3}H]-thymidine labeling of secondary glial cultures was 25% greater in cells from 24-month-old-versus three-month-old-donors (Fig. 2).

Fig. 2. Proliferation of astrocytes and microglia during primary culture from brains of ages across the life span (from Rozovsky et al. 1998)

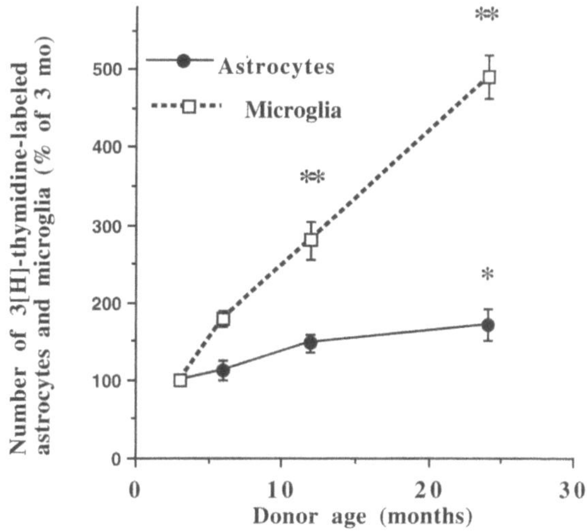

Although TGF-β1 inhibited proliferation of astrocytes and microglia in glial cultures from three-month-old-donors, proliferation of cultured glia from 24-month-old-brain was not inhibited by TGF-β1: astrocytes (Fig. 3A) and microglia (Fig. 3B). Because microglia from 24-months-old-brains lose responsiveness to the antiproliferative properties of TGF-β1, we examined if other responses to TGF-β1 were modified. As an anti-inflammatory cytokine, TGF-β1 inhibits activation of microglia originated from neonatal brain. Using NO production by microglia as an indicator of activation, we examined whether TGF-β1 would inhibit lipopolysaccharide (LPS)-induced activation of microglia. LPS treatment increased NO production in cultures from all age groups. However, TGF-β1 inhibited the LPS-induced increase in NO production only in microglia from three-month-old, but not 12 or 24-month-old-brains (Fig 4 A).

The age-related increase in number of activated microglia in vivo is consistent with the observed loss of sensitivity to TGF-β1. Further studies may reveal common mechanisms in the desensitization of responses to TGF-β1 in these different cell types. These age-related changes may contribute to neurodegenerative diseases in which activated microglia are hypothesized to promote neurodegeneration. We hypothesize that the observed activation of microglia at middle age could be a trigger in a slow cascade that predisposes the onset of AD and Parkinson's disease, which typically occur during middle age or later.

Because of age-related increases of GFAP mRNA and transcription (e.g. Nichols et al. 1993; Morgan et al. 1997, we compared the activity of the 1.9 kb of 5′-upstream region of the rat GFAP promoter, which has a rich array of physiological response elements (Laping et al. 1994; Rozovsky et al. 1998), including those that mediate responses to TGF-β1 and IL-1 (Krohn et al. 1999). In secondary monotypic astrocyte cultures originated from 3, 12,

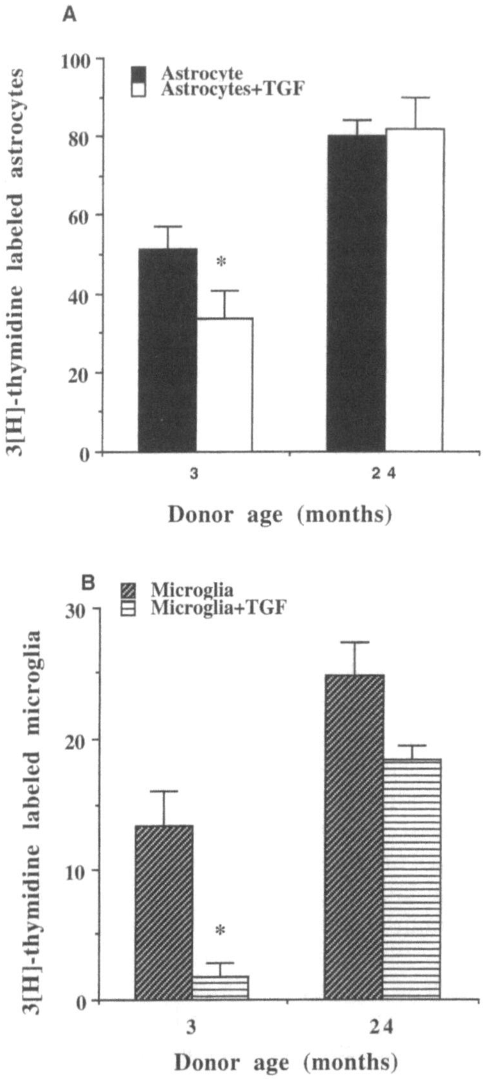

Fig. 3. Transforming growth factor beta-1 (TGF-β1) inhibits the proliferation of astrocytes (**A**) and microglia (**B**) from young but not older brains (redrawn from Rozovsky et al. 1998)

and 24-month-old-brains, the level of GFAP transcription was two-fold higher in astrocyte cultures from 12 and 24-month-old vs. three-month-old-donors (Fig. 4B). These findings indicate that the transcriptional factor inventory characteristic of astrocytes from aging brains is retained in primary culture, which gives a valuable opportunity for detailed analysis of the promoter sequences that mediate the increased transcription during aging and for the subsequent construction of transgenic mice with modified astrocyte aging patterns.

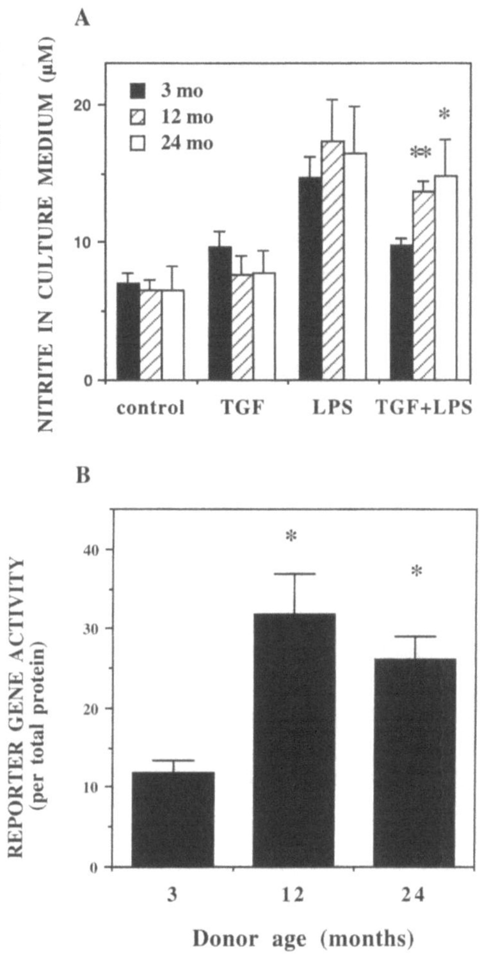

Fig. 4. **A** Transforming growth factor beta-1 (TGF-β1) inhibits lipopolysaccharide (LPS)-stimulated NO production in microglia from young but not older donors (from Rozovsky et al. 1998). **B** Primary astrocytes show that age increased transcription of the GFAP promoter with a luciferase reporter, as observed in vivo (from Rozovsky et al. 1998)

Implications for Proliferative Disorders of Brain Cells

The strikingly greater proliferation of astrocytes and microglia from aging brains is in opposition to that of cells from non-neural tissues. For example, primary cultures of human skin fibroblasts and vascular endothelial cells from humans and mice progressively lost proliferative capacity as a function of donor age (Martin et al. 1970, 1983). Although the greatest decrease occurs during postnatal maturation, further decreases are found during aging in adult donors to a degree that depends on the tissue source.

These findings of enhanced proliferation of glia from aging brains suggest the hypothesis that aging promotes a proliferative microenvironment in the brain. On one hand, normal human brains show an increased incidence of "tortuosity" and increased length and tortuosity of small blood vessels (e.g., Spangler et al. 1994) that are strongly suggestive of continued capillary endothelial growth. Brain vascular endothelia thus merits study for controls on proliferation as a function of adult age – e.g., does the diminished anti-proliferative effect of TGF-β1 extend to other brain cell types? Herein may be found important clues to age-related cerebrovascular changes, which are concurrent with endothelial thinning, such as the marked PAS reactivity that is indicative of glycoxidized extracellular matrix (Sobin et al. 1992) and cerebrovascular deposits of Aβ peptides. These phenomena may also be factors in the age-related increase of gliomas observed in human populations. Astrocytomas are rarely reported in aging laboratory rodents (an exception is described in Frith and Chandra 1990) and are not recognized as an age-related brain lesion in the common laboratory rodent genotypes. In human populations, the age-specific rate of brain tumors increases progressively from about 3/100,000 in young adults to more than 20/100,000 by age 70 (Jellinger 1978; Ahsan et al. 1995). However, this relatively low level of incidence is compatible with the rare findings of astrocytomas in aging rats, because the rodent populations sampled are usually groups of <100 animals. In humans, the most common glial cell tumor type may be glioblastoma multiforme, which typically has astrocytic features.

Manipulations of Glial Hyperactivity by Food Restriction

Because of the implied roles of activated glia in neurodegeneration in AD, e.g., as producers of free radicals in response to amyloid β-peptide (Aβ), it is of interest that glial aging can be manipulated by metabolic interventions. We found that chronic food restriction (FR) attenuates age-associated glial activation in the hippocampus, basal ganglia and corpus callosum (Morgan et al. 1997, 1999). Markers of astrocyte and microglial activation were compared in ad libitum fed (AL) rats age three and 24 month and chronic FR rats aged 24 months. Age-associated activation of astrocytes and microglia was attenuated by FR in most regions. We hypothesize that age-related glial activation is an inflammatory response to local oxidative stress and that FR decreases the load of oxidized macromolecules.

To model these complex in vivo processes, we are examining astrocyte-microglial interactions in culture. GFAP transcription is increased by reactive oxygen, in mixed or monotypic glia (Morgan et al. 1997). Thus, astrocyte activation may be independent from microglial activation. Interestly exogenous apoJ activates microglia (Xie et al. 1997) but is also neuroprotective in cultured hippocampal slices against ADDLs (Lanzrein et al. 1997). We hypothesize that activation of astrocytes during aging is an early event in a pathogenic cascade of AD that, by increasing apoJ secretion, also increases ADDL production (Lambert et al. 1998), leading to neuron damage. The wild-card in this game of sudden cell death may be fluctuations in microglial activation, which are induced by apoJ itself and by Aβ in various states of aggregation. These models may help evaluate anti-inflammatory interventions into AD.

Acknowledgments. These studies were supported by grants to CEF from the NIA and to TEM from The Alzheimer Association.

References

Ahsan H, Neugut AI, Bruce JN (1995) Trends in incidence of primary malignant brain tumors in USA, 1981–1990. Int J Epidemiol 24:1078–1085

Bons N, Jallageas V, Mestre-Frances N, Sihol S, Petter A, Delacourte A (1995) *Microcebus murinus;* a convenient laboratory animal model for the study of Alzheimer's disease. Alzheimer's Res. 1:83–87

Cork LC, Masters C, Beyreuther K, Price D (1990) Development of senile plaques: relationships of neuronal abnormalities and amyloid deposits. Am J Pathol 137:1383–1392.

Farkas I, Baranyi L, Takahashi M, Fukuda A, Liposits Z, Yamamoto T, Okada H (1998) A neuronal C5a receptor and an associated apoptotic signal transduction pathway. J Physiol (Lond) 507:679–87

Finch CE (1990) Longevity, senescence, and the genome. Chicago, Ill, University of Chicago Press.

Finch CE (1994) The evolution of ovarian oocyte decline with aging and possible relationships to Down syndrome and Alzheimer disease. Exp Gerontol 29:299–304.

Finch CE, Marchalonis J (1996) An evolutionary perspective on amyloid and inflammatory features of Alzheimer disease. Neurobiol. Aging 17:809–815

Finch CE, Sapolsky RM. (1999). The reproductive schedule apoE, and the evolution of Alzheimer disease. Neurobiol. Aging, in press

Frith CH, Chandra M (1990) Spontaneous primary astrocytoma in the spinal cord of a mouse and a rat. Toxicol Pathol 18:427–429

Gearing M, Rebeck WG, Hyman BT, Tigges J, Mirra SS (1994) Neuropathology and apolipoprotein E profile of aged chimpanzees: implications for Alzheimer disease. Proc Nat Acad Sci USA 91:9382–9386

Geula C, Wu CK, Saroff D, Lorenzo A, Yuan M, Yankner BA (1998) Aging renders the brain vulnerable to amyloid beta-protein neurotoxicity. Nat Med 4:827–831

Goldsmith SK, Wals P, Rozovsky I, Morgan TE, Finch CE (1997) Kainic acid and decorticating lesions stimulate the synthesis of C1q protein in adult rat brain. J Neurochem 68:2046–2052

Jellinger K (1978) Glioblastoma multiforme: morphology and biology. Acta Neurochir (Wien) 42:5–32

Jucker M, Walker LC, Martin LJ, Kitt, CA, Kleinman HK, Ingram DK, Price DL. (1992) Age-associated inclusions in normal and transgenic mouse brain. Science 255:1443–145

Krohn K, Rozovsky I, Wals P, Teter B, Anderson CP, Finch CE (1999) GFAP transcriptional responses to TGF-β1 and IL-1β in rat astrocytes are mediated by an NF-1 like site in the near-upstream promoter. J Neurochem 72:1353–1361

Lambert MP, Barlow AK, Chromy B, Edwards C, Freed R, Liosatos M, Morgan TE, Rozovsky I, Trommer B, Viola KL, Wals P, Zhang C, Finch CE, Krafft GA, Klein WL (1998) Diffusible, non-fibrillar ligands derived from $A\beta_{1-42}$. Proc Natl Acad Sci USA 95:6448 – 6453

Laping NJ, Teter B, Nichols NR, Rozovsky I, Finch CE (1994) Glial fibrillary acidic protein: regulation by hormones, cytokines, and growth factors. Brain Pathol 4:259–274

Lampert-Etchells M, Pasinetti GM, Finch CE, Johnson SA (1993) Regional localization of cells containing C1qb and C4 mRNAs in the frontal cortex during Alzheimer disease. Neurodegeneration 2:111–121

Lanzrein A-S, Finch CE, Morgan TE (1997) Protection of glutamate toxicity by clusterin in organotypic hippocampal slices. Soc Neurosci 23:1388

Loeffler DA, Brickman CM, Juneau PL, Perry MF, Pomara N, Lewitt PA (1997) Cerebrospinal fluid C3a increases with age, but does not increase further in Alzheimer's disease. Neurobiol Aging 18:555–557

Martin G, Sprague C, Epstein C (1970) Replicative life-span of cultivated human cells: Effects of donor's age, tissue, and genotype. Lab Invest 23:86–91

Martin GM, Obgurm CE, Wight TN (1983) Comparative rates of decline in the primary cloning efficiencies of smooth muscle cells from the aging thoracic aorta of two murine species of contrasting maximum life span potentials. Am J Pathol 110:236–245
McKee AC, Kowall NW, Schumacher JS, Beal MF (1998) The neurotoxicity of amyloid beta protein in aged primates. Amyloid 5:1–9
Morgan TE, Rozovsky I, Goldsmith SK, Stone DJ, Yoshida T, Finch CE (1997) Increased transcription of the astrocyte gene GFAP during middle-age is attenuated by food restriction: implications for the role of oxidative stress. Free Rad Biol Metab 23:524–528
Morgan TE, Xie Z, Goldsmith S, Yoshida T, Lanzrein A-S, Stone D, Rozovsky I, Perrv G, Smith MA, Finch CE (1999) The mosaic of brain glial hyperactivity during normal aging and its attenuation by food restriction. Neuroscience 89:687–699
Nichols NR, Day JR, Laping NJ, Johnson SA, Finch CE (1993) GFAP mRNA increases with age in rat and human brain. Neurobiol Aging 14:421–429
O'Banion K, Finch CE (1996) Editors' Introduction. Inflammatory mechanisms in Alzheimer disease: overview of this volume and a short history of the subject. Neurobiol Aging 17:669–671
Oda T, Wals P, Osterburg HH, Johnson SA, Pasinetti GM, Morgan TE, Rozovsky I, Stine WB, Snyder SW, Holzman TF, Krafft GA, Finch CE (1995) Clusterin (apoJ) alters the aggregation of amyloidβ-peptide ($A\beta_{1-42}$) and forms slowly sedimenting $A\beta$ complexes that cause oxidative stress. Exp Neurol 136:22–31
Pasinetti GM, Tocco G, Sakhi S, Musleh WD, DeSimoni M, Mascarucci P, Schreiber S, Baudry M, Finch CE (1996) Hereditary deficiencies in complement C5 are associated with intensified excitotoxic brain lesions that implicate new roles of the C-system in neuronal and astrocytic functions. Neurobiol Dis 3:197–205
Pasinetti GM, Hassler M, Stone D, Finch CE (1999) Glial gene expression during aging in rat striatum and in long-term responses to 6-OHDA lesions. Synapse 31:278–284
Perry VH, Matyszak M, Fearn S (1993) Altered antigen expression of microglia in aged rodent CNS. GLIA 7:60–67
Rogers J, Luber-Narod J, Styren SD, Civin WH (1988) Expression of immune system-associated antigens by cells of the human central nervous system: Relationship to the pathology of Alzheimer's disease. Neurobiol Aging 9:339–349
Rogers J, Cooper NR, Webster S, Schultz J, McGeer PL, Styren SD, Civin WH, Brachova L, Bradt B, Ward P, Lieberburg I (1992) Complement activation by β-amyloid in Alzheimer disease. Proc Nat Acad Sci USA 89:1016–1020
Rozovsky I, Morgan TE, Willoughby DA, Dugich-Djordevich MN, Pasinetti GM, Johnson SA, Finch CE (1994) Selective expression of clusterin (SG-2) and complement C1q and C4 during responses to neurotoxins in vivo and in vitro. Neuroscience 62:741–758
Rozovsky I, Finch CE, Morgan TE (1998) Age-related activation of microglia and astrocytes: in vitro studies show persistence of phenotypes of aging, increased proliferation, and resistance to down-regulation. Neurobiol Aging 19:97–103
Scott SA, Mandybur TI (1996) Astrocytic and microglial alterations in the aged mouse brain. In: U. Mohr (Ed.) Pathobiology of the aging mouse. Vol. 2. ISLI Press, Washington, pp. 39 – 52
Shen Y, Li R, McGeer EG, MeGeer PL (1997) Neuronal expression of mRNAs for complement proteins of the classical pathway in Alzheimer brain. Brain Res 769:391–395
Singhrao SK, Morgan BP, Neal JW, New GR (1995) A functional role for corpora amylacea based on evidence from complement studies. Neurodegeneration 4:335–345
Sobin S, Bernick S, Ballard KW (1992) Histochemical characterization of the aging microvasculature in the human and other mammalian and non-mammalian vertebrates by the periodic acid-Schiff reaction. Mech Age Devel 63:183–192
Spangler KM, Challa VR, Moody DM, Bell MA (1994) Arteriolar tortuosity of the white matter in aging and hypertension. A microradiographic study. J Neuropathol Exp Neurol 53:22–26
Velazquez P, Cribbs DH, Poulos TL, Tenner AJ (1997) Aspartate residue 7 in amyloid beta-protein is critical for classical complement pathway activation: implications for Alzheimer's disease pathogenesis. Nat Med 3:77–79
Xie Z, Wals PA, Finch CE, Goldsmith SK, Morgan TE (1997) Clusterin activates cultured mouse microglia. Soc Neurosci 23:2430

Mechanistic and Metaphorical Connections Between NF-κB and the Secreted Alzheimer's β-Amyloid Precursor Protein

S. W. Barger, X. Mao, A.M. Moerman, and A. Ranganathan

Summary

NF-κB is a designation originally referring to a DNA-binding activity that bound the κB enhancer of immunoglobulin κ light chain and stimulated transcription. The proteins responsible for this activity have since been cloned and studied extensively by immunologists, structural biochemists, and cell biologists. Only recently have neurobiologists turned their attention to NF-κB to study its role in the nervous system. Predictable roles for NF-κB in the regulation of glial cytokines have been confirmed. Less obvious roles have been suggested for NF-κB in processes peculiar to neurons, such as synaptic plasticity. Its participation in inflammation and its induction by toxic stimuli such as tumor necrosis factor (TNF) led to the common interpretation of NF-κB as a pathogenic factor. An alternative interpretation, that NF-κB is involved in compensatory responses to pathogens, was supported by our demonstration that some neuroprotective factors can activate NF-κB or a similar activity in neurons. At least one of these factors – secreted β-amyloid precursor protein (sAPP) – can also induce NF-κB and inflammatory reactions in microglia, the phagocytic inflammatory cells of the central nervous system. This fact makes sAPP a factor with both beneficial and harmful effects on neuronal health. This "double-edged" phenomenon and the factors that influence the balance of sAPP's effects are discussed herein.

The Biology of the β-Amyloid Precursor Protein

One of the pathognomic features of Alzheimer's disease is the neuritic (or "senile") amyloid plaque. In 1987, several laboratories cloned a sequence encoding the large protein from which is generated the amyloid β-peptide (Aβ; Goldgaber et al. 1987; Robakis et al. 1987; Tanzi et al. 1987), the major component of senile plaques. βAPP was found to be a glycoprotein expressed in at least four isoforms generated through differential splicing. The proteins all have a single membrane-spanning region, and some splice variants (e.g., βAPP751 and βAPP770) contain a Kunitz protease inhibitor (KPI) domain. Indeed, a portion of βAPP751 had been isolated previously and termed protease nexin II by virtue of its activity as a serine protease inhibitor (Van

Patterson/Kordon/Christen (Eds.)
Neuro-Immune Interactions in Neurologic and Psychiatric Disorders

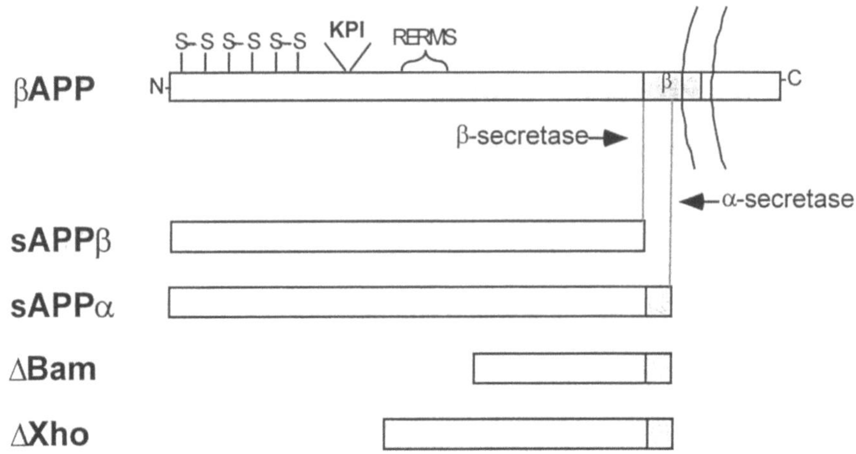

Fig. 1. sAPP structures. *β*APP (top) is synthesized as a transmembrane protein. A*β* is shaded. In the aminoterminus is a cysteine-rich region (S-S), followed by a site for potential insertion of the KPI in splice variants, and the "RERMS" region. The large extracellular portion can be shed from cells by either the *α*- or *β*-secretase. Below these natural products are depicted the artificial deletion mutants ΔBam and ΔXho, distinguished by their respective exclusion or inclusion of the "RERMS" region

Nostrand and Cunningham 1987). In addition to the proteolytic cleavage event (*β*-secretase cleavage) that initiates generation of A*β*, *β*APP can undergo an alternative proteolysis (*α*-secretase cleavage) that liberates most of the extracellular portion of *β*APP as a diffusible, extracellular protein designated sAPP*α* (Fig. 1). The *α*-secretase event has been the subject of intense study because it abrogates the generation of A*β*, a popular candidate for the common pathogenic mediator of Alzheimer's. Studies conducted by Nitsch and coworkers 1992) determined that cholinergic stimuli can elevate *α*-secretase processing of *β*APP. This phenomenon appears to involve muscarinic activation of protein kinase C by the phosphatidylinositol hydrolysis pathway, as direct stimulation of these signal transduction events can have similar effects (Buxbaum and Greengard 1996). The *β*-secretase efficiency can be altered as well, most notably by mutations in the *β*APP sequence that segregate genetically with rare forms of familial Alzheimer's disease (Haass et al. 1995). Data suggest that *β*-secretase cleavage of *β*APP can occur in several different locations within the cell and can lead to an alternative secreted product (sAPP*β*). For a more complete review of the proteolytic processing of *β*APP, see Hardy (1997).

Biological Effects of sAPP*α*

The controlled release of sAPP from neurons and other cell types suggests that this species has some defined biological role. The secreted portions of

βAPP751 and βAPP770 probably play important roles in controlling protease activity. However, βAPP695 – the neuron-specific isoform – does not contain the KPI domain. Several laboratories therefore embarked on an exploration of the potential functions of the sAPPα derived from βAPP695. (All the studies described in this chapter were performed with this splice variant. Therefore, use of "sAPPα" or "sAPPβ" hereafter will refer to the products of βAPP695). Prolific work was conducted by Saitoh and coworkers, arising from the discovery that sAPPα could be mitogenic for fibroblasts (Saitoh et al. 1989). Saitoh's laboratory went on to show that this activity, as well as a neurite-extension activity, depended on a 17-amino acid sequence containing the core sequence R-E-R-M-S (Jin et al. 1994). However, sAPPα is a structurally complex molecule with several other structural motifs that would appear to have activities by virtue of their similarity to other proteins. At least two heparin-binding sites have been identified in the extracellular portion of βAPP, and these may play roles in cell attachment and neuritogenesis (Jin et al. 1994; Small et al. 1994); it is unclear what role these heparin-binding regions might play in the secreted moieties. In the aminoterminal quarter of sAPPα is a cysteine-rich region that has been connected to the activation of the Erk mitogen-activated kinases through a Ras-dependent mechanism (Greenberg et al. 1994).

It was eventually discovered that sAPPα has an ability to protect central nervous system neurons from several toxic treatments. This neuroprotective activity is effective against excitotoxic glutamate exposure, the metabolic stress of glucose deprivation, and oxidative stress evoked by iron or Aβ itself (Barger et al. 1995a; Goodman and Mattson 1994; Mattson et al. 1993). In some cases, the higher survival rate of neurons treated with sAPPα probably results from the ability of the protein to moderate or depress intracellular ionic calcium concentrations ($[Ca^{2+}]_i$) rapidly through an elevation of cyclic GMP levels (Barger et al. 1995a). This idea is consistent with the well-characterized contribution of $[Ca^{2+}]_i$ elevations to neurotoxicity in several paradigms (Mattson and Barger 1993). The elevation of cGMP by sAPPα is connected to $[Ca^{2+}]_i$ depression through a hyperpolarization caused by the activation of a potassium current (Furukawa et al. 1996a). In addition to enhanced survival rates in the face of excitotoxic challenges, sAPPα-evoked changes in membrane potential can have potentially meaningful physiological consequences with respect to synaptic efficacy (Ishida et al. 1997; Mucke et al. 1994; Roch et al. 1994).

sAPPα and κB-Binding Transcription Factors in Neurons

Because the elevation of cGMP and hyperpolarization evoked by sAPPα are extremely rapid, the protein can alleviate excitotoxicity when applied simultaneously with glutamate. However, survival rates of neurons pretreated with sAPPα for several hours are even higher than those seen in cells receiving sAPPα coincident with the toxin. This finding suggests that sAPPα effects a long-lasting change in the cell, perhaps through changes in gene expression. As an initial exploration of the gene-regulatory influence of sAPPα, we assayed κB-binding transcription factors. The responsiveness of NF-κB to cellular oxidation had suggested to some investigators that it was a participant in the death-promoting effects of free radicals. We worked under the contrary assumption that NF-κB was a compensatory response factor that is involved in battling oxidative damage. If so, it might also participate in programs of gene expression that are activated by neuroprotective agents. Therefore, we tested the ability of sAPPα to enhance transcription from a reporter gene construct that contained κB elements in its promoter. Both sAPPα and a cGMP analog activated this reporter system in transfected neuroblastoma cell lines, and the effect of sAPPα was blocked by an antagonist of the cGMP-dependent kinase (Barger and Mattson 1996a,b). In primary hippocampal neurons, we monitored the activity of κB-binding factors by electrophoretic mobility shift assay (EMSA), which tests for activity in nuclear extracts that can bind DNA of a specific sequence. Secreted APPα and cGMP induced a κB-binding activity in this model as well (Barger and Mattson 1996a). We generally have abstained from referring to this activity as "NF-κB" because it has characteristics inconsistent with the factor classically referred to by this term (below).

To test more specifically the influence of κB-dependent transcription on neuronal survival, we sought to activate NF-κB more directly. Latent NF-κB resides in the cytosol, held in a latent state by an inhibitory subunit generally termed IκB. Activation results from the targeted degradation of IκB by the ubiquitin-proteosome system (Desterro et al. 1998; Palombella et al. 1994). We treated primary neurons and neuroblastoma cells with antisense oligonucleotides directed against one of the major IκB proteins, IκBα. These treatments apparently lowered the translation rate of IκBα and led to a transient induction of a κB-binding activity (Barger and Mattson 1996a). In neurons containing presumptive NF-κB activated in this way, survival rates were higher in the face of challenges with calcium ionophore, excitotoxic levels of glutamate, and Aβ (Barger et al. 1995b; Barger and Mattson 1996 a, b).

Since those initial studies, elevation of κB-dependent transcription has been correlated with neuroprotection in other situations. TNF has biological effects similar to sAPPα (although it does not appear to elevate cGMP or evoke rapid depressions of $[Ca^{2+}]_i$). TNF activates κB-dependent transcription in neurons and provides protection against glutamate, Aβ, and ferrous sulfate (which generates free radical peroxide via the Fenton reaction) (Barger 1997; Barger et al. 1995b; Barger and Mattson 1996a). Ceramide, an endogenous signal transduction molecule, also can activate NF-κB and promote neuronal survival in the presence of toxins (Goodman and Mattson 1996). Finally, we found similar effects for dehydroepiandrosterone-sulfate (DHEA-S; Mao and Barger 1998), a humoral and brain-produced steroid that declines during aging and is inversely correlated with Alzheimer's disease (Nasman et al. 1991). In the case of DHEA-S, blockade of the κB-binding activity abrogated the neuroprotective effects of the steroid.

Oxidative stress is a common component of the toxic treatments against which κB-dependent transcription was protective, and it was assumed that NF-κB serves to elevate antioxidant enzyme expression, as it would be expected to do in compensatory responses to reactive oxygen species. This assumption has been confirmed recently in studies that documented an elevation of manganese (mitochondrial) superoxide dismutase in neurons by NF-κB (Mattson et al. 1997). However, treatments that elevate neuronal κB-binding activity also squelch $[Ca^{2+}]_i$, particularly the elevations of $[Ca^{2+}]_i$ evoked by glutamate (Barger et al. 1995b; Barger and Mattson 1996a). A portion of this effect may involve an elevated expression of the calcium-binding protein calbindin. Calbindin has been connected to enhanced resistance to calcium-dependent insults (Meier et al. 1998) and is induced by TNF (Mattson et al. 1995). Potential influences of κB-dependent transcription on the expression of glutamate receptor subunits or other proteins involved in the activity of glutamate receptors remain to be investigated. Most recently, intracellular proteins that act to inhibit apoptotic mechanisms have been implicated in the survival-promoting effects of NF-κB (Wang et al. 1998; Wu et al. 1998).

NF-κB has been reported to respond to ionotropic glutamate receptor stimulation (Grilli et al. 1996; Guerrini et al. 1994, 1995; Kaltschmidt et al. 1995). However, several of these studies relied on immunocytochemical detection of NF-κB, which is not tantamount to DNA-binding activity. In those studies that demonstrated increased DNA binding through EMSA, the cultures were typically grown in the presence of serum, which can allow the proliferation of glial cells. Because activation of neuronal NF-κB by glutamate should exert a neuroprotective influence, we were interested in clarifying this effect. We tested the influence of glutamate receptor stimulation on κB binding in cultures of nearly pure cortical neurons (99.33%). In these cultures, a predominant constitutive κB-binding factor is detected. This activity is distinguished from the conventional NF-κB (a heterodimer of RelA and p50 protein subunits) by antigenicity, DNA target preference, and electrophoretic mobility under both native and denaturing conditions (Moerman et al. 1999). We have tentatively termed this factor *n*euronal κB-*b*inding *f*actor (NKBF). This factor is inducible by a cGMP analog, but glutamate diminishes its activity (Mao et al. 1999). In contrast, cocultures of primary neurons with glia show induction of *bona fide* NF-κB by glutamate. Use of a stratified coculture arrangement that allows for independent harvest of neurons and glia demonstrates that the NF-κB induced by glutamate in cocultures is localized to the glia. Interestingly, glia cultured in the absence of neurons do not respond to glutamate. The influence of glutamate on NKBF in neurons indicates that delayed excitotoxicity may involve suppression of the neuroprotection otherwise offerred by κB-dependent transcription factors.

Effects of sAPPα on Micrologia

The activation of κB-dependent transcription by sAPP would have dramatically different effects on the brain if localized to microglia. NF-κB induction is generally associated with inflammatory responses, and proinflammatory activation of microglia can lead to the production of specific neurotoxins, as well as generalized increases in free radical production and pleiotrophic cytokines (Breitner 1996). Thus, its induction of κB-binding activity raises the possibility that sAPP, an otherwise neuroprotective agent, could mediate indirect neurotoxicity through activation of NF-κB in microglia. The first tenet of this hypothesis was tested through EMSA of nuclear extracts from microglial cells. At nanomolar concentrations, sAPP was found to induce a κB-binding activity that was identified as classical NF-κB by reactivity with antibodies to RelA and p50 (Barger and Harmon 1997). Unlike the induction of NKBF, activation of microglial NF-κB by sAPP is independent of cGMP.

Several gene products have been identified that participate in microglial proinflammatory reactions through their induction by NF-κB. An inducible isoform of nitric oxide synthase (iNOS) is one of the genes controlled by NF-κB, especially in rodent cells and tissues. This enzyme catalyzes the breakdown of arginine into citrulline, thereby releasing NO. Although NO can be toxic, it can also serve a neuroprotective role, depending largely on its oxidation state (Lipton et al. 1993). In addition, the human iNOS gene appears to be less responsive to NF-κB than its rodent counterparts (Colton et al. 1996). Therefore, the contribution of iNOS and NO to inflammatory neurodegenerative disease is controversial. Nevertheless, the enzyme and its chemical products can serve as a useful index of microglia activation. When further oxidized (primarily by superoxide, another product of activated microglia), NO forms nitrite; this can be detected through a simple colorimetric assay known as the Griess reaction (e.g., Grisham et al. 1996). Secreted APP was found to elevate both iNOS protein levels and nitrite concentration in rodent microglia (Barger and Harmon 1997). Aβ also has been shown to activate microglia if applied with a costimulus such as interferon-γ (Meda et al. 1995). However, sAPP is approximately 1000-fold more potent and over six times as efficacious as Aβ in the stimulation of nitrite production (Fig. 2).

NF-κB also controls the expression of several cytokines. Interleukin-1 (IL-1) is a class of particular interest to Alzheimer's disease researchers. The discovery of microglia overexpressing IL-1 in Alzheimer's (Griffin et al. 1989) was one of the watershed events in the indictment of inflammatory mechanisms in this disease. Both IL-1α and IL-1β have diverse biological influences on local tissues and systemically (Griffin et al. 1998). IL-1 can elevate the expression of βAPP in CNS neurons (Buxbaum et al. 1992; Goldgaber et al. 1989), suggesting that it could contribute to the deposition of Aβ. It can also interfere with long-term potentiation of synapses (Murray and Lynch, 1998), an effect that could contribute to memory dysfunctions. We found that the effects of sAPP on microglia include an elevation of IL-1β production

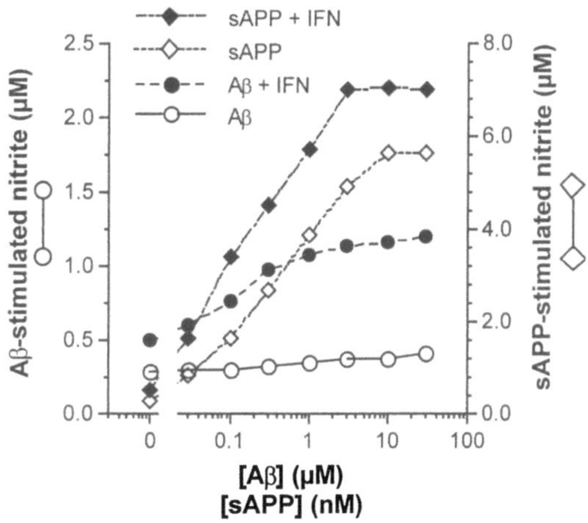

Fig. 2. Comparison of the microglial activation by sAPP and Aβ. Primary microglia were treated with the indicated concentrations of sAPPα (nM) (diamonds) or Aβ_{1-42} (μM) (circles). These agonists were applied in the absence (open symbols) or presence (filled symbols) of 10 ng/ml interferon-γ. After 20 hours, nitrite was measured in the culture medium by Griess reaction. Data reflect the mean of quadruplicate determinations. Note that the responses to Aβ are scaled on the left ordinate and sAPPα responses on the right

(Barger and Harmon 1997). This finding illustrates a potential vicious circle, through which IL-1 and βAPP contribute to one another's overexpression. The effects of sAPP on microglia in general indicate a potential contributor to the inflammatory events apparent in Alzheimer's disease, and this scenario is consistent with the preferential association of activated microglia with plaques containing dystrophic neurites overexpressing βAPP (Cras et al. 1990; Griffin et al. 1995).

Structural Determinants of sAPP Activities

The generation of Aβ from βAPP involves a β-secretase event, a proteolytic cleavage distinct from that which produces sAPPα. The large soluble product of β-secretase activity (sAPPβ) is shorter by 15 or 16 amino acids on its carboxyterminus (Anderson et al. 1991; Esch et al. 1990). To explore the implications of increased β-secretase processing for sAPP biology, we compared the activities of sAPPα and sAPPβ in several paradigms. Sequence coding for human sAPP was placed in a bacterial expression vector with a "STOP" codon engineered after amino acid 612 (sAPPα) or 596 (sAPPβ). After purification, these proteins were tested for the ability to protect neurons from the rapid phase of toxicity mediated by glucose deprivation.

Neurons cultured in the absence of glucose die in large percentages within 20 hours. After a 24-hour pretreatment with 10 nM sAPPα, significantly fewer neurons died; however, sAPPβ was deficient in this activity (Fig. 3).

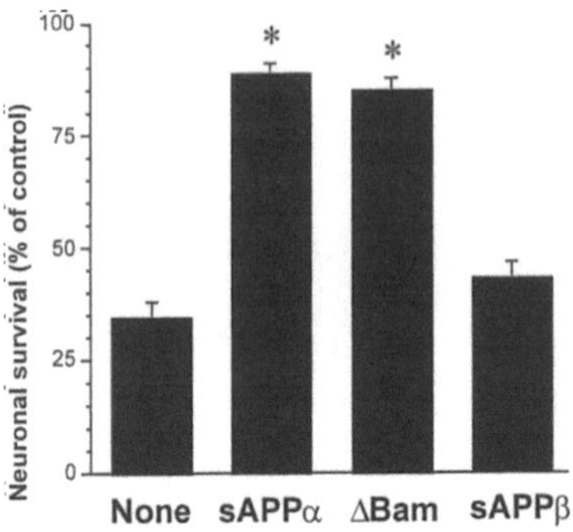

Fig. 3. Comparison of sAPPα and sAPPβ in a neuroprotective assay. Primary hippocampal neurons were pretreated for 24 hours with 10 nM sAPPα, sAPPβ, ΔBam, or no protein (None). Photomicrographs were then taken of random fields, and the culture medium was replaced by one lacking glucose. After 18 hours photomicrographs were taken of the same fields and surviving neurons (identified by established morphological criteria) were counted. Values reflect the percentage of survival as a mean of triplicate cultures ± SEM. (*$p<0.01$ vs. None")

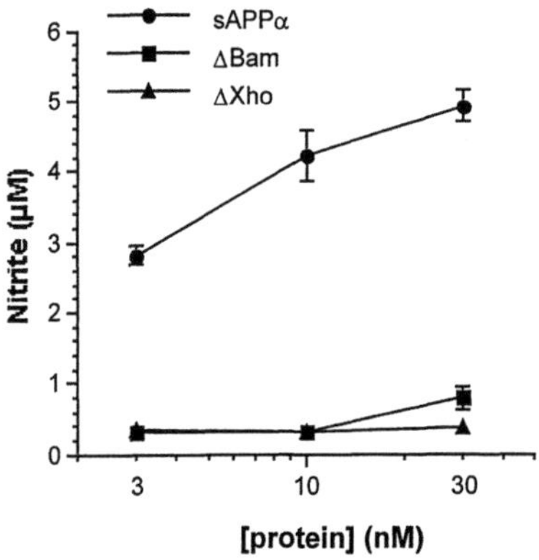

Fig. 4. Test of the role of the "RERMS" region in the proinflammatory activity of sAPP. Primary microglia were treated with the indicated concentrations of sAPPα, ΔBam, or ΔXho (illustrated in Fig. 1). After 2o hours, nitrite was measured in the culture medium by Griess reaction. Data reflect the mean ± SEM of quadruplicate determinations

This finding is consistent with comparisons of sAPPα and sAPPβ in other tests of neuromodulation (Furukawa et al. 1996b; Li et al. 1997). However, the results were different with respect to microglial activation: both sAPPα and sAPPβ were able to stimulate nitrite production by microglia with similar potency and efficacy (Barger and Harmon, 1997). Therefore, it appears that the neuroprotective activity and the proinflammatory activity of sAPP can be attributed to structurally distinct elements, consistent with the differential involvement of cGMP in the activation of κB-binding factors in neurons versus glia. One implication of these data is that increased β-secretase processing of βAPP would generate a protein that can stimulate proinflammatory reactions without providing neuroprotection.

The ability of sAPPβ to activate microglia suggests that this activity results from residues aminoterminal to the Aβ region. As mentioned above, one domain previously connected to bioactivity is a 17-amino acid sequence including the critical residues R-E-R-M-S. We tested the involvement of this region through deletion mutants in which it is alternatively included or excluded (Fig. 1). Deletion of aminoterminal residues of sAPPα severely diminished proinflammatory activity, and the "RERMS" region did not affect this deficiency (Fig. 4). Thus, the ability of sAPP to activate microglia requires residues in the extreme aminoterminus, analogous to the structural requirements previously ascribed to the activation of Erk kinases (Greenberg et al. 1994).

Modulation by Apolipoprotein E

While mutations in the βAPP gene are responsible for a small fraction of Alzheimer's cases, allelic differences in the gene for apolipoprotein E (apoE) are more important demographically (Roses 1997). In most regions of the world, approximately 15% of the general population possesses an $\varepsilon 4$ allele of apoE. But this factor approximately doubles the risk for Alzheimer's disease before the age of 75, compared to those possessing only the more common $\varepsilon 3$ allele; therefore, the frequency of the $\varepsilon 4$ allele approaches 30% in the Alzheimer population. Homozygotes for $\varepsilon 4$ face an even greater risk. Recent analyses illustrate that the apoE allele does not dictate whether an individual will develop Alzheimer's but instead influences the age of onset (Meyer et al. 1998).

Considering that the βAPP and apoE loci can influence Alzheimer's individually, we wondered whether this could be indicative of a functional relationship between the encoded proteins. The simplest possibility is that the proteins might interact with one another physically. To test this idea, we performed coprecipitation tests of sAPPα:ApoE mixtures. In this paradigm, an anti-ApoE antibody is able to precipitate a complex that includes sAPPα (Fig. 5; Barger and Harmon 1997; Barger and Mattson 1997). Likewise, anti-APP antibody precipitates proteins including ApoE. We also performed precipitation by an alternative means: sAPPα synthesized with a polyhistidine tag allows it to be precipitated with a nickel-chelating resin. This system also demonstrates coprecipitation of ApoE when the proteins are mixed (Barger and Harmon 1997; Barger and Mattson 1997). Comparisons between mixtures involving the protein products of the $\varepsilon 3$ and $\varepsilon 4$ apoE gene indicate that ApoE3 binds sAPPα more efficiently than does ApoE4. Studies with deletion constructs of sAPPα demonstrated a requirement for the aminoterminal amino acid residues (Barger and Harmon 1997; Barger and Mattson 1997). This binding and its structural determinates were confirmed by Hass et al. (1998), who found that the aminoterminal 207 residues of APP were sufficient for binding to ApoE.

Present data indictate that the structural requirements of sAPPα for ApoE binding are identical to those for microglial activation. Therefore, we compared the actions of sAPPα on microglia to those of the sAPPα/ApoE complex. As might be predicted, ApoE interfered with the ability of sAPPα to activate microglia (Barger and Harmon 1997). Consistent with their relative

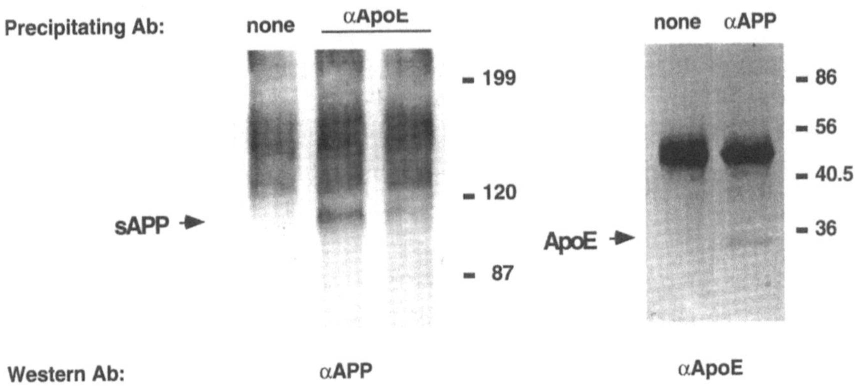

Fig. 5. ApoE and sAPP coprecipitation. sAPP*α* was incubated alone (left panel, lane 3) or with ApoE3 (all other lanes); the concentration of each protein was 450 nM. After two hours, the mixtures were precipitated with protein-A and antibody against ApoE or APP or with protein-A alone ("none"). The immunoprecipitates were subjected to western blot analysis using the antibody complementary to that used in precipitation; positions of molecular weight markers are indicated at the right of each panel (kDa). The bands corresponding to sAPP*α* or ApoE are designated; other bands are contaminants from the protein-A preparation

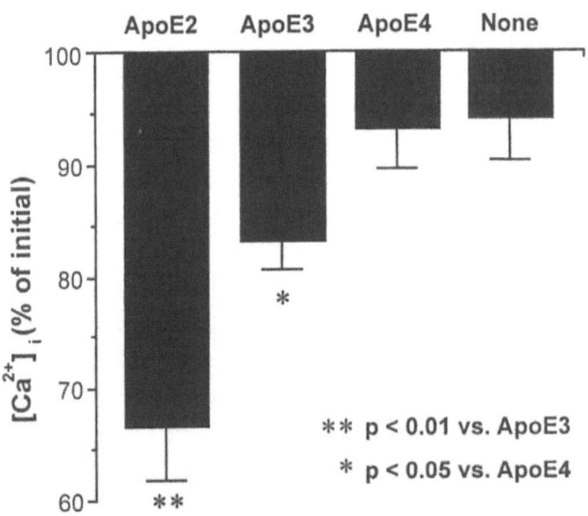

Fig. 6. Relative modulation of sAPP activity by isoforms of ApoE. sAPP*α* was incubated with the indicated protein (or alone: "None") for 60 minutes; the concentration of each protein was 50 nM. The mixture was then applied to hippocampal neurons at a final concentration of 50 pM (each protein). Depression of $[Ca^{2+}]_i$ in the neurons at 200 seconds after treatment was recorded and is expressed as a percentage of the value before treatment. Each point represents the mean ± SEM from 32–37 cells in three separate cultures

binding efficiencies, ApoE3 was more potent in this respect than was ApoE4. To the extent that the proinflammatory activity of sAPP might participate in the initiation or progression of Alzheimer's, apoE genotype could thus influence these mechanisms. In support of this idea, the degree of microglial activation in Alzheimer brains is modulated by apoE genotype (Egensperger et al. 1998).

Because the neuromodulatory/neuroprotective effects of sAPPα appear to depend on a portion of the molecule distinct from that which mediates microglial activation, it might be predicted that ApoE binding would have no influence on the former activity. However, coincubation with ApoE actually potentiates the neuromodulatory activity of sAPPα (Barger and Mattson 1997). The effect is dependent upon the time and concentration parameters during the coincubation, suggesting that it depends upon the physical interaction between the two proteins rather than an independent effect of ApoE on the cells. Again, there is a distinction between ApoE3 and ApoE4, with the latter being less potent (Barger and Mattson 1997). More recent data extend this genetic correlation to the $\varepsilon 2$ allele product, as well (Fig. 6). Studies have found that the $\varepsilon 2$ allele confers some degree of protection against developing Alzheimer's (Corder et al. 1994), and our data demonstrate that its protein product is even more efficient than ApoE3 at modulating sAPPα activity.

sAPP and NF-κB: Neuronal Life in the Balance

Together, the data presented here suggest a multi-faceted model for the role of sAPP in Alzheimer's disease and related conditions. It is evident that APP expression is elevated in discrete cellular compartments in Alzheimer's, particularly in the dystrophic neurites surrounding senile plaques (Cras et al. 1991; Joachim et al. 1991; Shoji et al. 1990). Other conditions that can lead to Alzheimer's disease or Alzheimer-related pathology also involve elevation of βAPP, including epilepsy, stroke, and head trauma. It is likely that this induction is part of a compensatory response, relying on the direct neuroprotective activity of sAPPα. When the protein binds to vulnerable neurons, it facilitates a protective expression of antioxidant enzymes and other proteins through the activation of an NF-κB-like transcription factor. ApoE, also elevated by conditions of stress or injury, binds to and interferes with the region of sAPPα capable of stimulating inflammatory reactions. However, several situations may interfere with this beneficial scenario: 1) the elevation of sAPPα may outpace the levels of ApoE, releasing the proinflammatory region of the molecule from its inhibition; 2) an individual could express ApoE4, an isoform less effective at inhibiting proinflammatory actions of sAPPα; 3) an individual could possess mutations in βAPP that have been demonstrated to generate a higher sAPPβ:sAPPα ratio; there may be metabolic abnormalities that would shift βAPP processing in this way as well; 4) finally, microglial numbers or activation state may be enhanced by other inflammatory stimuli, magnifying the consequences of sAPP's bioactivity. The result of either would be activation of NF-κB in microglia, triggering the production of neurotoxins that cause synaptic dysfunction, structural damage, and neuronal death (Fig. 7). Preliminary evidence suggests that a substantial fraction of

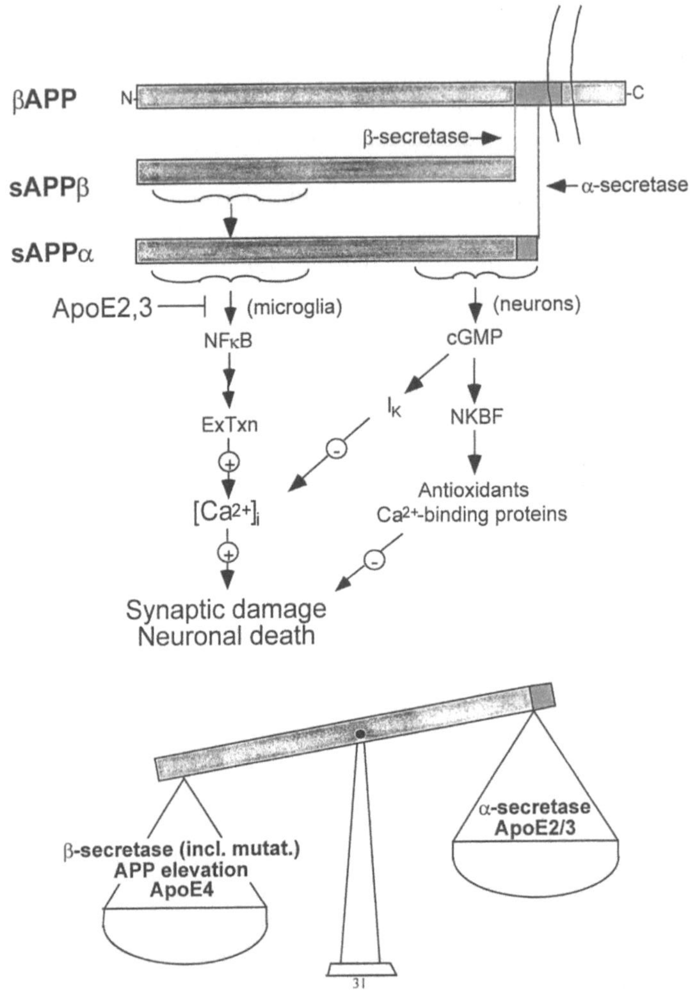

Fig. 7. The sAPP bioactivity balance model. βAPP and its alternative sAPP products are illustrated. The N-terminal region of sAPP stimulates the left side of the bioactivity flow chart in microglia, ultimately leading to the production of an excitotoxin (ExTxn) capable of harming the function or viability of neurons. These events are inhibited by binding of ApoE2 or ApoE3. These events also are antagonized by the neuroprotection afforded through the C-terminal extension present in sAPPα (which operates through cGMP-dependent activation of a potassium current [1_K] and activation of NKBF). Mutations in βAPP and physiological stimuli can promote β-secretase activity, ApoE4 genotype is permissive for greater microglial activation, and APP elevation may outpace ApoE levels; all these latter events would tip the balance toward synaptic and neuronal degeneration

the neurotoxicity manifest in sAPP-activated microglia is excitotoxic in nature; this finding is consistent with the findings of others (Giulian et al. 1996; Lipton 1994) and reinforces the importance of calcium regulation in these events.

In the model presented above, the integrated effect of sAPP on neuronal function and viability depends on a delicate balance between direct, beneficial effects on neurons and indirect, harmful effects through microglia. Through their control of both neuroprotective and inflammation-related genes, NF-κB and related transcription factors are similar double-edged swords. Indeed, its responsiveness to a wide variety of stimuli makes NF-κB potentially more significant in this tenuous balance. Therefore, as therapeutic strategies are sought among the increasing number of anti-inflammatory drugs, care must be taken to account for the complexity of the role of NF-κB in neuronal life and death.

References

Anderson JP, Esch FS, Keim PS, Sambamurti K, Lieberburg I, Robakis NK (1991) Exact cleavage site of Alzheimer amyloid precursor in neuronal PC-12 cells. Neurosci Lett 128:126–128

Barger SW (1997) Tumor necrosis factor: the good, the bad, and the umbra. In: Mattson MP (ed) Neuroprotective Signal Transduction. Humana, Totawa, NJ, pp. 163–183

Barger SW, Mattson MP (1996a) Induction of neuroprotective κB-dependent transcription by secreted form of the Alzheimer's β-amyloid precusor. Mol Brain Res 40:116–126

Barger SW, Mattson MP (1996b) Participation of gene expression in the protection against amyloid β-peptide toxicity by the β-amyloid precursor protein. Ann NY Acad Sci 777:303–308

Barger SW, Mattson MP (1997) Isoform-specific modulation by apolipoprotein E of the activities of secreted β-amyloid precursor protein. J Neurochem 69:60–66

Barger SW, Harmon AD (1997) Microglial activation by secreted Alzheimer amyloid precursor protein and modulation by apolipoprotein E. Nature 388:878–881

Barger SW, Fiscus RR, Ruth P, Hofmann F, Mattson MP (1995a) The role of cyclic GMP in the regulation of neuronal calcium and survival by secreted forms of β-amyloid precursor. J Neurochem 64:2087–2096

Barger SW, Hörster D, Furukawa K, Goodman Y, Kriegelstein J, Mattson MP (1995b) Tumor necrosis factors α and β protect neurons against amyloid β-peptide toxicity: Evidence for involvement of a κB-binding factor and attenuation of peroxide and Ca^{2+} accumulation. Proc Natl Acad Sci USA 92:9328–9332

Breitner JCS (1996) The role of anti-inflammatory drugs in the prevention and treatment of Alzheimer's disease. Annu Rev Med 47:401–411

Buxbaum JD, Greengard P (1996) Regulation of APP processing by intra- and intercellular signals. Ann N Y Acad Sci 777:327–331

Buxbaum JD, Oishi M, Chen HI, Pinkas-Kramarski R, Jaffe EA, Gandy SB, Greengard P (1992) Cholinergic agonists and interleukin 1 regulate processing and secretion of the Alzheimer β/A4 amyloid protein precursor. Proc Natl Acad Sci USA 89:10075–10078

Colton C, Wilt S, Gilbert D, Chernyshev O, Snell J, Dubois-Dalcq M (1996) Species differences in the generation of reactive oxygen species by microglia. Mol Chem Neuropathol 28:15–20

Corder EH, Saunders AM, Risch NJ, Strittmatter WJ, Schmechel DE, Gaskell PCJ, Rimmler JB, Locke PA, Conneally PM, Schmader KE (1994) Protective effect of apolipoprotein B type 2 allele for late onset Alzheimer disease. Nat Genet 7:180–184

Cras P, Kawai M, Siedlak S, Mulvihill P, Gambetti P, Lowery D, Gonzalez-DeWhitt P, Greenberg B, Perry G (1990) Neuronal and microglial involvement in β-amyloid protein deposition in Alzheimer's disease. Am J Pathol 137:241–246

Cras P, Kawai M, Lowery D, Gonzaiez-DeWhitt P, Greenberg B, Perry G (1991) Senile plaque neurites in Alzheimer disease accumulate amyloid precursor protein. Proc Natl Acad Sci USA 88:7552–7556

Desterro JMP, Rodriguez MS, Hay RT (1998) SUMO-1 modification of IκBα inhibits NF-κB activation. Mol Cell 2:233–239

Egensperger R, Kosel S, von Eitzen U, Graeber MB (1998) Microglial activation in Alzheimer disease: Association with APOE genotype. Brain Pathol 8:439–447

Esch FS, Kiem PS, Beattie BC, Blacher RW, Culwell AR, Olsterdorf T, McClure D, Ward PJ (1990) Cleavage of amyloid β peptide during constitutive processing of its precursor. Science 248:1122–1124

Furukawa K, Barger SW, Blalock EM, Mattson MP (1996a) Activation of potassium channels and suppression of neuronal activity by secreted β-amyloid precursor protein. Nature 379:74–78

Furukawa K, Sopher BL, Rydel RE, Begley JG, Pham DG, Martin GM, Fox M, Mattson MP (1996b) Increased activity-regulating and neuroprotective efficacy of α-secretase-derived secreted amyloid precursor protein conferred by a c-terminal heparin-binding domain. J Neurochem 67: 1882–1896

Giulian D, Yu J, Li X, Tom D, Li J, Wendt E, Lin SN, Schwarcz R, Noonan C (1996) Study of receptor-mediated neurotoxins released by HIV-1-infected mononuclear phagocytes found in human brain. J Neurosci 16:3139–3153

Goldgaber D, Lerman MI, McBride OW, Saffiotti U, Gajdusek DC (1987) Characterization and chromosomal localization of a cDNA encoding brain amyioid of Alzheimer's disease. Science 235:877–880

Goldgaber D, Harris HW, Hla T, Maciag T, Donnelly RJ, Jacobsen JS, Vitek MP, Gajdusek DC (1989) Interleukin-1 regulates synthesis of amyloid β-protein precursor mRNA in human endothelial cells. Proc Natl Acad Sci USA 86:7606–7610

Goodman Y, Mattson MP (1994) Secreted forms of β amyloid precursor protein protect hippocampal neurons against β-peptide toxicity and oxidative injury. Exp Neurol 128:1–12

Goodman Y, Mattson MP (1996) Ceramide protects hippocampal neurons against excitotoxic and oxidative insults, and amyloid β-peptide toxicity. J Neurochem 66:869–872

Greenberg SM, Koo EH, Selkoe DJ, Qiu WQ, Kosik KS (1994) Secreted β-amyloid precursor protein stimulates mitogen-activated protein kinase and enhances tau phosphorylation. Proc Natl Acad Sci USA 91:7104–7108

Griffin WS, Sheng JG, Roberts GW, Mrak RE (1995) Interleukin-1 expression in different plaque types in Alzheimer's disease: significance in plaque evolution. J Neuropathol Exp Neurol 54:276–281

Griffin WST, Stanley LC, Ling C, White L, MacLeod V, Perrot LJ, White CLI, Araoz C (1989) Brain interleukin 1 and S-100 immunoreactivity are elevated in Down syndrome and Alzheimer disease. Proc Natl Acad Sci USA 86:7611–7615

Griffin WS, Sheng JG, Royston MC, Gentleman SM, McKenzie JE, Graham DI, Roberts GW, Mrak RE (1998) Glial-neuronal interactions in Alzheimer's disease: the potential role of a 'cytokine cycle' in disease progression. Brain Pathol 8:65–72

Grilli M, Goffi F, Memo M, Spano P (1996) Interleukin-1 β and glutamate activate the NF-κB/Rel binding site from the regulatory region of the amyloid precursor protein gene in primary neuronal cultures. J Biol Chem 271:15002–15007

Grisham MB, Johnson GG, Lancaster JR, Jr. (1996) Quantitation of nitrate and nitrite in extracellular fluids. Methods Enzymol 268:237–246

Guerrini L, Blasi F, Denis-Donini S (1994) Activation of NFκB upon glutamate stimulation in cultured mouse cerebellar granule cells. Soc Neurosci Abstr 20:1510

Guerrini L, Blasi F, Denis-Donini S (1995) Synaptic activation of NF-κB by glutamate in cerebellar granule neurons in vitro. Proc Natl Acad Sci USA 92:9077–9081

Haass C, Lemere CA, Capell A, Citron M, Seubert P, Schenk D, Lannfelt L, Selkoe DJ (1995) The Swedish mutation causes early-onset Alzheimer's disease by β-secretase cleavage within the secretory pathway. Nature Med 1:1291–1296

Hardy J (1997) Amyloid, the presenilins and Alzheimer's disease. Trends Neurosci 20:154–159

Hass S, Fresser F, Kochl S, Beyreuther K, Utermann G, Baier G (1998) Physical interaction of ApoE with amyloid precursor protein independent of the amyloid Aβ region in vitro. J Biol Chem 273:13892–13897.

Ishida A, Furukawa K, Keller JN, Mattson MP (1997) Secreted form of β-amyloid precursor protein shifts the frequency dependency for induction of LTD, and enhances LTP in hippocampal slices. Neuroreport 8:2133–2137

Jin LW, Ninomiya H, Roch JM, Schubert D, Masliah E, Otero DA, Saitoh T (1994) Peptides containing the RERMS sequence of amyloid β/A4 protein precursor bind cell surface and promote neurite extension. J Neurosci 14:5461–5470

Joachim C, Games D, Morris J, Ward P, Frenkel D, Selkoe D (1991) Antibodies to non-β regions of the β-amyioid precursor protein detect a subset of senile plaques. Am J Pathol 138:373–384

Kaltschmidt C, Kaltschmidt B, Baeuerle PA (1995) Stimulation of ionotropic glutamate receptors activates transcription factor NF-κB in primary neurons. Proc Natl Acad Sci USA 92:9618–9622

Li HL, Roch JM, Sundsmo M, Otero D, Sisodia S, Thomas R, Saitoh T (1997) Defective neurite extension is caused by a mutation in amyloid β/A4 (Aβ) protein precursor found in familial Alzheimer's disease J Neurobiol 32:469–480

Lipton SA (1994) AIDS-related dementia and calcium homeostasis. Ann N Y Acad Sci 747:205–224

Lipton SA, Choi YB, Pan ZH, Lei SZ, Chen HS, Sucher NJ, Loscalzo J, Singel DJ, Stamler JS (1993) A redox-based mechanism for the neuroprotective and neurodestructive effects of nitric oxide and related nitroso-compounds. Nature 364:626 – 632

Mao X, Barger SW (1998) Neuroprotection by dehydroepiandrosterone-sulfate: involvement of an NFκB-like factor. NeuroReport 9:759–763

Mao X, Moerman AM, Lucas MM, Barger SW (1999) Inhibition of a neuronal κB-binding factor by glutamate. J Neurochem (in press)

Mattson MP, Barger SW (1993) Roles for calcium signalling in structural plasticity and pathology in the hippocampal system. Hippocampus 3:73–87

Mattson MP, Cheng B, Culwell AR, Esch FS, Lieberburg 1, Rydel RE (1993) Evidence for excitoprotective and intraneuronal calcium-regulating roles for secreted forms of the β-amyloid precursor protein. Neuron 10:243–254

Mattson MP, Cheng B, Baldwin SA, Smith-Swintosky VL, Keller J, Geddes JW, Scheff SW, Christakos S (1995) Brain injury and tumor necrosis factors induce calbindin D-28k in astrocytes: evidence for a cytoprotective response. J Neurosci Res 42:357–370

Mattson MP, Goodman Y, Luo H, Fu W, Furukawa K (1997) Activation of NF-κB protects hippocampal neurons against oxidative stress-induced apoptosis: evidence for induction of manganese superoxide dismutase and suppression of peroxynitrite production and protein tyrosine nitration. J Neurosci Res 49:681–697

Meda L, Cassatella MA, Szendrei GI, Otvos L, Jr., Baron P, Villalba M, Ferran D, Rossi F (1995) Activation of microglial cells by β-amyloid protein and interferon-γ. Nature 374:647 – 650

Meier TJ, Ho DY, Park TS, Sapolsky RM (1998) Gene transfer of calbindin D28k cDNA via herpes simplex virus amplicon vector decreases cytoplasmic calcium ion response and enhances neuronal survival following glutamatergic challenge but not following cyanide. J Neurochem 71: 1013 – 1023

Meyer MR, Tschanz JT, Norton MC, Welsh-Bohmer KA, Steffens DC, Wyse BW, Breitner JC (1998) APOE genotype predicts when-not whether-one is predisposed to develop Alzheimer disease. Nat Genet 19:321–322

Moerman AM, Mao X, Lucas MM, Barger SW (1999) Characterization of a neuronal κB-binding factor distinct from NF-κB. Mol Brain Res 67:303–315

Mucke L, Masliah E, Johnson WB, Ruppe MD, Alford M, Rockenstein EM, Forss-Petter S, Pietropaolo M, Mallory M, Abraham CR (1994) Synaptotrophic effects of human amyloid β protein precursors in the cortex of transgenic mice. Brain Res 666:151–167

Murray CA, Lynch MA (1998) Evidence that increased hippocampal expression of the cytokine interleukin-1β is a common trigger for age- and stress-induced impairments in long-term potentiation. J Neurosci 18:2974–2981

Nasman B, Olsson T, Backstrom T, Eriksson S, Grankvist K, Vutanen M, Bucht G (1991) Serum dehydroepiandrosterone sulfate in Alzheimer's disease and in multi-infarct dementia. Biol Psych 30:684–690

Nitsch RM, Slack BE, Wurtman RJ, Growdon JH (1992) Release of Alzheimer amyloid precursor derivatives stimulated by activation of muscarinic acetylcholine receptors. Science 258:304–307

Palombella VJ, Rando OJ, Goldberg AL, Maniatis T (1994) The ubiquitin-proteasome pathway is required for processing the NF-κB1 precursor protein and the activation of NF-κB. Cell 78:773–785

Robakis NK, Ramakrishna N, Wolfe G, Wisniewski HM (1987) Molecular cloning and characterization of a cDNA encoding the cerebrovascular and the neuritic plaque amyloid peptides. Proc Natl Acad Sci USA 84:4190–4194

Roch JM, Masliah E, Roch-Levecq AC, Sundsmo MP, Otero DA, Veinbergs I, Saitoh T (1994) Increase of synaptic density and memory retention by a peptide representing the trophic domain of the amyloid β/A4 protein precursor. Proc Natl Acad Sci USA 91:7450–7454

Roses AD (1997) Apolipoprotein E, a gene with complex biological interactions in the aging brain. Neurobiol Dis 4:170–185

Saitoh T, Sundsmo M, Roch J-M, Ximura M, Cole G, Schubert D, Oltersdorf T, Schenk DB (1989) Secreted form of amyloid β protein precursor is involved in the growth regulation of fibroblasts. Cell 58:615–622.

Shoji M, Hirai S, Yamaguchi H, Harigaya Y, Kawarabayashi T (1990) Amyloid β-protein precursor accumulates in dystrophic neurites of senile plaques in Alzheimer-type dementia. Brain Res 512:164–168

Small DH, Nurcombe V, Reed G, Clarris H, Moir R, Beyreuther K, Masters CL (1994) A heparin-binding domain in the amyloid protein precursor of Alzheimer's disease is involved in the regulation of neurite outgrowth. J Neurosci 14:2117–2127

Tanzi RE, Gusella JF, Watkins PC, Bruns GA, St George-Hyslop P, Van Keuren ML, Patterson D, Pagan S, Kurnit DM, Neve RL (1987) Amyloid β protein gene: cDNA, mRNA distribution, and genetic linkage near the Alzheimer locus. Science 235:880–884

Van Nostrand WE, Cunningham DD (1987) Purification of protease nexin II from human fibroblasts. J Biol Chem 262:8508–8514

Wang CY, Mayo MW, Korneluk RG, Goeddel DV, Baldwin AS, Jr. (1998) NF-κB-antiapoptosis: induction of TRAF1 and TRAF2 and c-IAP1 and c-IAP2 to suppress caspase-8 activation. Science 281:1680–1683

Wu MX, Ao Z, Prasad KV, Wu R, Schlossman SF (1998) IEX-1L, an apoptosis inhibitor involved in NF-κB-mediated cell survival. Science 281:998–1001

Pathological and Biochemical Studies of Chronic Neuroinflammation May Lead to Therapies for Alzheimer's Disease

G.L. Wenk, B. Hauss-Wegrzyniak, and L.B. Willard

Introduction

The discovery of an effective pharmacotherapy for the prevention or treatment of Alzheimer's Disease (AD) will ultimately depend upon the development of an appropriate animal model that reproduces the multiple components of the disease.

AD is characterized by a pattern of neuropathological changes, such as *β*-amyloid plaques, forebrain cholinergic deficiency, impaired memonic abilities, and a dense distribution of activated astrocytes and microglia (McGeer et al. 1984; Rinne et al. 1987; Swaab et al. 1994; Terry et al. 1991; Vogels et al. 1990). A decline in the number of cholinergic cells in the basal forebrain (BF) may underlie the cognitive impairments (Terry et al. 1991). The mechanisms underlying the degeneration of BF cholinergic cells are unknown. Chronic inflammatory processes may play an important role in the etiology or pathogenesis of AD (Aisen and Davis 1994; McGeer et al. 1989; Mrak et al. 1995; Rogers 1995). Many biomarkers associated with inflammatory activity are elevated in AD brains. For example, AD is associated with increased levels of the inflammatory cytokine interleukin-1 in brain tissue (Griffin et al. 1989). Reactive gliosis is a prominent component of neuritic plaques (Griffin et al. 1989). Reactive glia and elevations in inflammatory cytokines may contribute to the pathology seen in AD (Frederickson and Brunden 1994; Griffin et al. 1989).

The series of studies described herein investigated the usefulness of a novel animal model of AD that has evolved from an increased appreciation of the potential role of chronic neuroinflammation in the progression of AD. We have shown previously that this animal model reproduces the following components of AD: 1) increased numbers of activated astrocytes and microglia, particularly within the hippocampus, entorhinal cortex and basal forebrain; 2) increased levels of interleukin-1*α* and *β*, apolipoprotein E, tumor necrosis factor-*α* and *β*-amyloid precursor protein mRNA in hippocampus and basal forebrain; 3) hippocampal pathology associated with a significant impairment in working memory; and 4) a time-dependent, but not dose-dependent, death of basal forebrain cholinergic cells (Hauss-Wegrzyniak et al. 1998a; Willard et al. 1999).

Using this animal model, we investigated 1) the ability of two drugs to attenuate the pathological, biochemical and behavioral effects of chronic neu-

Patterson/Kordon/Christen (Eds.)
Neuro-Immune Interactions in Neurologic and Psychiatric Disorders

roinflammation, and 2) the time-dependent evolution of the distribution of activated microglia. The results suggest that a combined therapy using a non-steroidal anti-inflammatory drug and an antagonist of glutamate receptors might significantly attenuate the progression or symptoms of AD. Most importantly, the results obtained using this novel animal model suggest that the selective vulnerability of temporal lobe regions as well as the cholinergic neurons within the nucleus basalis of Meynert in humans might be due to the inflammatory processes that are active in the brains of AD patients.

In the present series of studies we investigated the biochemical, neuropathological and behavioral consequences of chronic brain infusion of a pro-inflammatory agent, lipopolysaccharide (LPS) into the basal forebrain (BF) or fourth ventricle of young rats.

Studies of Chronic Inflammation Within the Basal Forebrain

We have previously shown that a chronic, low dose LPS infusion into the BF can destroy, possibly selectively, cholinergic cells (Willard et al. 1999). We investigated whether the chronic administration of a non-competitive N-methyl-D-aspartate-sensitive (NMDA) glutamate antagonist, memantine, could provide significant neuroprotection from the cytotoxic effects of LPS-induced neuroinflammation.

Osmotic minipumps (Alzet, model 2004) containing LPS (Sigma, *E. coli*, serotype 055:B5, TCA extraction, 1.0 μg/μl) were implanted into the dorsal abdomen and attached via tubing to a chronic indwelling cannula positioned stereotaxically so that the cannula tip extended into the BF at the following coordinates: AP: -0.6 mm, ML + 2.8 mm, and DV: -7.9 mm (from the skull). Controls were infused with artificial cerebrospinal fluid (aCSF) vehicle. The minipumps delivered 0.25 μl/hr of either LPS or aCSF for 37 days.

The chronic LPS infusions decreased neocortical choline acetyltransferase (ChAT; Fonnum 1969) activity, decreased the number of ChAT-immunoreactive cells within the BF that send afferents to the neocortex, and increased the number of activated microglia. Daily peripheral administration of memantine (20 mg/kg/day, i.p.) significantly attenuated the cytotoxic effects of the chronic inflammation upon BF cholinergic cells. These results suggest that the cytotoxic effects of chronic neuroinflammation may operate via NMDA receptors.

Studies of Chronic Global Neuroinflammation

We also investigated whether a cyclooxygenase (COX) inhibitor, nitroflurbiprofen (NFP, 15 mg/kg, s.c.; Nicox, France), could reduce the inflammatory response induced by continuous infusion (30 days by osmotic minipump as described above) of LPS into the fourth ventricle of young rats. We also

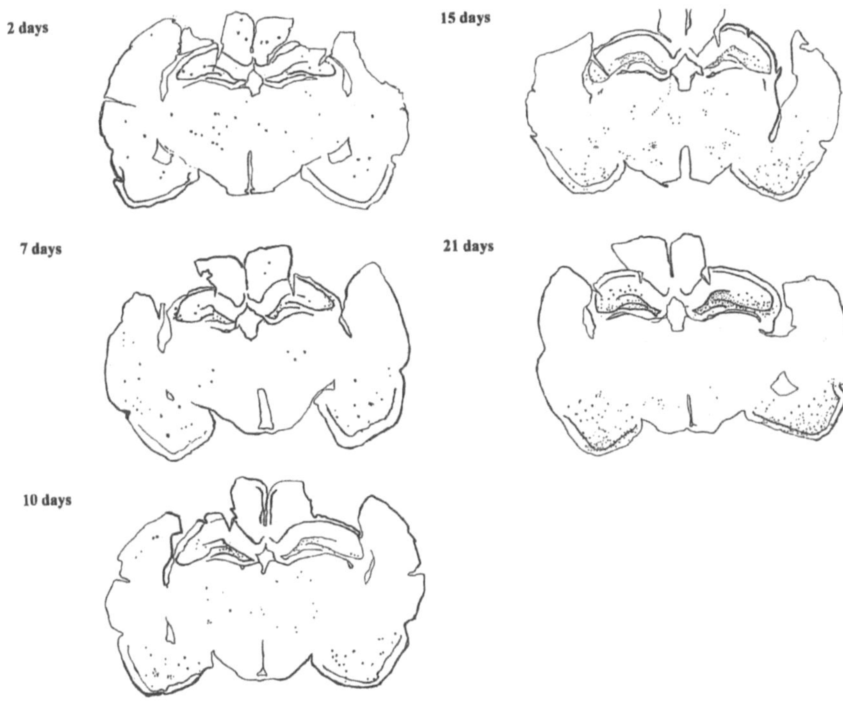

Fig. 1. Distribution of activated microglia within the brains of rats that were infused chronically with LPS for either 2, 7, 10, 15 or 21 days. The microglia were stained using the OX-6 antibody. Each dot represents a collection of five activated microglia. The dorsal regions of cortex were removed for biochemical analyses prior to the in situ perfusion

investigated the time-dependent evolution of the distribution and density of activated microglia that appear in response to the LPS infusion.

Rats were sacrificed at various times after the infusion and drug treatment began and were prepared for immunohistochemical analysis. The location and density of activated microglia were visualized using the monoclonal antibody OX-6 (final dilution 1:400; Pharmingen, San Diego) that is directed against the Class II major histocompatibility complex antigen (Hauss-Wegrzyniak et al. 1998b).

On day 2, immunohistochemical staining found activated microglia scattered fairly evenly throughout the brain (see Fig. 1). However, by day 21 the greatest inflammatory response following chronic exposure to LPS was contained only within the hippocampus, particularly the dentate gyrus, and within the entorhinal and piriform cortex. These results clearly show that temporal lobe regions respond differently, as compared to other brain regions, to the continued presence of an inflammogen such as LPS. The continued activation of microglia would lead to the long-term exposure of temporal lobe neurons to potentially toxic levels of cytokines and complement proteins. A similar condition might exist in the brains of AD patients in response to an appropriate inflammogen, such as elevated levels of soluble β-amyloid (Rogers at al. 1992). This could explain the apparent concentration of AD-related, pathology, i.e., senile plaques and tangles, within the temporal regions and the hippocampus.

In all LPS-treated rats the OX-6-positive microglia displayed a bushy morphology characterized by a contraction of their highly ramified processes and a large cell body. In LPS-infused rats treated with NFP at 15 mg/kg, the OX-6 antibody stained fewer ramified microglial cells, and these microglia were characterized by long, thin ramified processes; there were only a few bushy cells in these treated rats. Overall, the daily administration of NFP significantly attenuated the brain inflammation, as indicated by the decreased density and reactive state of microglial cells. The results demonstrated that NFP is an effective anti-inflammatory agent.

Taken together, these findings offer experimental support to recent epidemiological investigations that have reported a reduced risk of AD and a delayed progression of the disease in subjects treated chronically with anti-inflammatory drugs for rheumatic disorders. The results of these studies further suggest that AD patients in the early stages of their disease might significantly benefit from a combined chronic therapy of anti-inflammatories and selective, non-competitive NMDA receptor antagonists.

References

Aisen PS, Davis KL (1994) Inflammatory mechanisms in Alzheimer's disease: implications for therapy. Am J Psychiat 151:1105–1113

Fonnum F (1969) Radiochemical micro assays for the determination of choline acetyltransferase and acetylcholinesterase activities. Biochem J 115:465–472

Frederickson RCA, Brunden KR (1994) New opportunities in AD research – Roles of immunoinflammatory responses and glia. Alz Dis Assoc Dis 8:159–165

Griffin WST, Stanley LC, Ling C, White L, MacLeod V, Perrot LJ, White CL III, Arora C (1989) Brain interleukin-1 and S-100 immunoreactivity are elevated in Down Syndrome and Alzheimer disease. Proc Natl Acad Sci USA 86:7611–7615

Hauss-Wegrzyniak B, Dobranski P, Stoehr JD, Wenk GL (1998a) Chronic neuroinflammation in rats reproduces components of the neurobiology of Alzheimer's disease. Brain Res 780:294–303

Hauss-Wegrzyniak B, Lukovic L, Bigaud M, Stoeckel ME (1998b) Brain inflammatory response induced by intracerebroventricular infusion of lipopolysaccharide: an immunohistochemical study. Brain Res 794:211–224

McGeer PL, McGeer EG, Suzuki J, Dolman CE, Nagai T (1984) Aging, Alzheimer's disease and the cholinergic system of the basal forebrain. Neurology 34:741–745

McGeer PL, Akiyama H, Itagaki S, McGeer EG (1989) Immune system response in Alzheimer's disease. Can J Neurol Sci 16:516–527

Mrak RE, Sheng JG, Griffin WST (1995) Glial cytokines in Alzheimer's disease: review and pathogenic implications. Human Pathol 26:816–823

Rinne JO, Paljarvi L, Rinne K (1987) Neuronal size and density in the nucleus basalis of Meynert in Alzheimer's disease. J Neuronal Sci 79:67–76

Rogers J, Cooper N, Webster 5, Schultz J, McGeer PL, Styren SD, Civin WH, Brachova L, Bradt B, Ward P, Lieberburg I (1992) Complement activation by β-amyloid in Alzheimer's disease. Proc Natl Acad Sci USA 89:10016–10020

Rogers J (1995) Inflammation as a pathogenic mechanism in Alzheimer's disease. Drug Res 45:439–442

Swaab DF, Hofman MA, Lucassen PJ, Salehi A, Uylings HBM (1994) Neuronal atrophy, not cell death, is the main hallmark of Alzheimer's disease. Neurobiol Aging 15:369–371

Terry RD, Masliah E, Salmon DP, Butters N, DeTeresa R, Hill R, Hansen LA, Katzman R (1991) Physical basis of cognitive alterations in Alzheimer's disease: Synapse loss is the major correlate of cognitive impairment. Ann Neurol 30:572–580

Vogels OJM, Broere CAJ, Ter Laak HJ, Ten Donkelaar HJ, Nieuwenhys R, Schulte BPM (1990) Cell loss and shrinkage in the nucleus basalis Meynert complex in Alzheimer's disease. Neurobiol Aging, 11:3–13

Willard LB, Hauss-Wegrzyniak B, Wenk GL (1999) The pathological and biochemical consequences of acute and chronic neuroinflammation within the basal forebrain of rats. Neuroscience 88:193–200

Serotonin-Immune Interactions in Major Depression

M. Kubera and M. Maes

Introduction

A series of studies over the last decade has demonstrated that depression is associated with alterations in immune functions. Most of the initial research indicated that depression is accompanied by signs of in vitro immunosuppression, such as decreased mitogen-stimulated lymphocyte proliferation (Kronfol et al. 1983) and natural killer cell activity (Irwin et al. 1987). Recently, a large body of evidence has been reported suggesting that major depression is associated with activation of cell-mediated immunity (for reviews, see Smith 1991; Maes et al. 1995; Maes 1997). The evidence for this includes the following: activation of in vivo cell-mediated immunity, as indicated by 1) a significantly increased $CD4^+/CD8^+$ ratio, which is determined by an increased percentage of T helper ($CD4^+$) and T memory cells ($CD4^+CD45RO^+$) and/or by lowered percentage of T-suppressor cells ($CD8^+CD57$); 2) significantly increased numbers of activated T cells ($CD25^+$ and $HLA\text{-}DR^+$ T cells) in the peripheral blood and increased concentrations of serum soluble interleukin-2 receptor (sIL-2R); 3) enhanced production of mono- and T-lymphocytic products, such as interleukin-1β (IL-1β), IL-6 and interferon-γ(IFN-γ) by mitogen-stimulated peripheral blood mononuclear cells (PBMCs), as well as increased plasma concentrations of IL-6 and sIL-6R; and 4) elevated plasma and urine neopterin concentrations. Moreover, there is strong evidence that the acute episode of major depression is accompanied by activation of the inflammatory response system (IRS; for reviews, see Maes 1993, 1997). The evidence includes, among other things: increased numbers of peripheral blood leukocytes, neutrophils and monocytes; increased levels of prostaglandin E2 (PGE2), positive acute phase proteins (APPs), e.g., C-reactive protein, haptoglobin, alpha-1 acid glycoprotein as well as decreased concentrations of negative AP reactants, such as albumin and zinc; and increased serum elastase concentrations (for review, see Maes 1997).

At first sight, the findings on IRS activation in depression seem to contradict the finding that major depression is also accompanied by signs of immunosuppression (Kronfol et al. 1983; Irwin et al. 1987). However, immune responses are often accompanied by signs of immune activation as well as suppression, whereby a primary immune activation may be counterbalanced by multiple feedback systems that limit the initial immune response, e.g., hypersecretion of cytokines, cytokine receptors, cytokine receptor antagonists, pos-

Patterson/Kordon/Christen (Eds.)
Neuro-Immune Interactions in Neurologic and Psychiatric Disorders

itive APPs, glucocorticoids and catecholamines, and lowered serum concentrations of zinc. It has been hypothesized that increased secretion of proinflammatory cytokines, such as IL-1β, IL-6 and IFNγ, may underlie many, if not all, signs of the immune and AP response in that illness (for review, see Maes 1997). Indeed, it is well known that IL-1β and IL-6 synergize strongly in T-cell differentiation or proliferation and are major mediators of the AP response, whereas IFNγ activates the production of monocytic cytokines.

Moreover, animal literature strongly supports the contention that proinflammatory cytokines induce a "depressive-like" state and "sickness behavior". Treatment with high doses of cytokines, such as interferons, may induce behavioral changes, including depressive symptoms (for review, see Maier and Watkins 1998). Therefore, it has been hypothesized that proinflammatory cytokines play a role in generating the immune and AP response in depression and may contribute to the pathogenesis of major depression (Maes 1997).

Effects of Cytokines on Serotonin Metabolism

Immune constituents, such as cytokines, have profound effects on the serotonergic systems in the brain and in the periphery. Peripheral and central administration of IL-1β, IFNγ and TNFα significantly increase extracellular 5-HT concentrations in several brain areas such as the hypothalamus, the hippocampus, and the cortex (Clement et al. 1997). Intraperitoneal injection of LPS and IL-1 increased mouse brain concentrations of the 5-HT catabolite, 5-hydroxyindoleacetic acid (5-HIAA), and tryptophan in all brain regions examined (Dunn 1992).

Cytokines, such as IL-1, IFNγ, TNFα, not only acutely stimulate 5-HT neurotransmission but also reduce the production of 5-HT by stimulating the activity of indoleamine-2,3-dioxygenase (IDO), the first enzyme of the kynurenine pathway, which converts tryptophan, the precursor of 5-HT, to kynurenic acid and quinolinic acid (QUIN). QUIN is an endogenous N-methyl-D-aspartate (NMDA) receptor agonist. It has been shown that stimulation of NMDA receptors by QUIN can lead to neuronal damage and dysfunction. Activation of IDO causes an accumulation of other neurotoxic agents like L-kynurenine and 3-hydroxykynurenine. Accumulation of neuroactive tryptophan metabolites within the brain has a functional and clinical significance and can occur in a broad spectrum of inflammatory diseases. Overstimulation of IDO, which occurs in infections and inflammation, may be detrimental because it leads to depletion of the plasma concentrations of tryptophan and reduced synthesis of 5-HT in the brain, which depends on the plasma availability of tryptophan, with concomitant neurological effects like lowering of mood (Heyes et al. 1992). IRS activation may cause a reduced availability of tryptophan to the brain not only by inducing the tryptophan-catabolizing enzyme IDO, but also by using peripheral amino acids for leukocyte activation and synthesis of positive APPs in the liver.

Serotonin-Immune Interactions in Major Depression

There is converging evidence that disorders in the central and peripheral neurotransmission of serotonin (5-HT) are implicated in the pathogenesis or pathophysiology of major depression (Maes and Meltzer 1995) The primary serotonergic disorders in major depression are: 1) decreased activity of the presynaptic serotonergic neuron, which is related to a decreased availability of plasma tryptophan to the brain; 2) lowered activity of the 5-HT transporter; and 3) adaptive alterations in postsynaptic receptors, such 5-HT2A receptors (for review, see Maes and Meltzer 1995). Each of these serotonergic alterations may be influenced by activation of the IRS.

One of the most consistently reported abnormal findings in serotonergic metabolism in major depression is a reduced availability of plasma tryptophan to the brain (Maes and Meltzer 1995). Thus, serum tryptophan as well as the ratio of tryptophan to the sum of amino acids known to compete for the same cerebral uptake mechanisms (competing amino acids, e.g., tyrosine, valine, leucine, isoleucine, phenylalanine) is lower in major depressed subjects than in normal controls (for review, see Maes and Meltzer 1995). Total and free tryptophan as well as the tryptophan/competing amino acid ratio are indicators for the availability of tryptophan to the brain and, hence, for 5-HT synthesis in the brain.

Recently, it was suggested that, in depression, lower plasma tryptophan may be related to activation of the IRS. Clinical studies showed significant inverse relationships between various immune or inflammatory markers and plasma total tryptophan concentrations. For example, highly significant relationships were found between serum total tryptophan availability and the number of leukocytes, the $CD4^{+}/CD8^{+}$ T-cell ratio, plasma haptoglobin, IL-6 and neopterin (all negative) and total serum protein, serum transferrin, albumin, zinc, and (all positive; Maes et al. 1993, 1996, 1997b). Therefore, we have hypothesized that the lowered availability of tryptophan to the brain in patients with depression may be the consequence of IRS activation in that illness.

The tryptophan-immune interrelationships occur not only in major depression but also in chronic alcoholic patients without apparent liver disease (AWLD) and in patients with dementia (Maes et al. 1998; Bonaccorso et al. 1998). For example, AWLD patients have significantly lower plasma tryptophan and significantly higher serum IL-1RA and IL-8 concentrations than normal volunteers. In AWLD patients, there are significant negative correlations between the availability of tryptophan to the brain and serum IL-6, IL-8 and IL-1RA. The results suggest that lowered availability of plasma tryptophan to the brain in detoxified AWLD patients is related to activation of the IRS.

Some APPs, e.g., α1-acid glycoprotein and α1-antitrypsin, and immunoglobulin G (IgG) are endogenous, non-specific inhibitors of high-affinity [^{3}H]-imipramine binding to platelets (Strijewski et al. 1988). α1-Acid glycoprotein was confirmed to be the most potent plasma protein inhibitor

(Strijewski et al. 1988). Barkai et al. (1986a, b) have shown that α1-acid glycoprotein attenuates [^{3}H]-imipramine binding to cerebral membranes. Since plasma α1-acid glycoprotein is significantly increased in patients with major depression, this endogenous inhibitor of [^{3}H]-imipramine binding sites could play a role in 5-HT transporter dysfunctions in depression (Nemeroff et al. 1990).

Significant effects of proinflammatory cytokines on the 5-HT transporter system have been described. This plasma membrane receptor plays a central role in serotonergic neurotransmission by reuptake of 5-HT in the synaptic cleft and it is a primary target for antidepressants. IL-1β modulates the activity of the 5-HT transporter (Ramamoorthy et al. 1995). Tumor necrosis factor-α (TNFα) enhances 5-HT reuptake in the choriocarcinoma cell line JAR, whereas IL-6 has no effect. In human B cell lines, TNFα modulates 5-HT reuptake preferentially in cell lines homozygous for the long allele of the 5-HT transporter polymorphism.

In rats, it was found that lipopolysaccharide (LPS), a bacterial endotoxin which activates the production of several proinflammatory cytokines, markedly decreased DOI-induced wet dog shakes, suggesting that IRS activation affects 5-HT2A receptor-stimulated behaviors (Kouhata et al. 1998).

Another finding that links activation of the IRS to serotonergic disturbances in depression is the recent report by Sluzewska et al. (1997a) that serotonin antibodies are increased in major depression. During the acute phase of depression, there was a significant increase in antibody reactivity to both 5-HT and gangliosides. Antibodies against 5-HT and gangliosides were found to be significantly increased in treatment-resistant depressed (TRD) patients and coincided with lower serum 5-HT and the presence of IRS activation, including increased plasma IL-6 concentrations (Sluzewska et al. 1997a). It was hypothesized that increased production of IL-6 could be involved in the production of the above autoantibodies.

Effects of 5-HT on Immune Function

The involvement of the central serotonergic neurotransmitter system and peripheral 5-HT in the modulation of immune system activity is receiving increasing interest. Macrophages possess a specific active 5-HT uptake system similar in affinity to that of platelets. T lymphocytes constitutively express 5-HT receptors, such as 5-HT1A and 5-HT2A/2C receptors, as well as a high affinity 5-HT transporter. After stimulation with polyclonal activators or IFNγ, monocytes and T lymphocytes release 5-HT. The uptake of [^{3}H]-5-HT by lymphocytes is potently inhibited by antidepressants such as clomipramine, fluoxetine and fluvoxamine. Devoino et al. (1968) showed that different chemical substances, which have the common property of increasing 5-HT concentrations, inhibit immunogenesis and antibody production in rodents. These changes are related to central rather than peripheral 5-HT actions, because lesions of serotoninergic raphe nuclei promoted stimulation in the immune response. Devoino also showed that 5-HT suppresses some aspects of cell-mediated immunity such as delayed hypersensitivity and transplantation immunity (for review; see Devoino and Morozova 1988). Bonnet et al. (1984) reported that 5-HT suppresses murine lymphocyte response to phytohemagglutinin A and/or allogeneic cells. Other immune functions, however, may be stimulated by 5-HT. For example, cutaneous injection of 5-HT can initiate a delayed-type hypersensitivity reaction through local recruitment and activation of $CD4^+$ T-helper cells (Ptak et al. 1991). 5-HT enhances murine T lymphocyte proliferation in vitro; low doses (100 ng/ml) of exogenously added 5-HT stimulate T-cell proliferation in response to IL-2 (Young et al. 1993; Young and Matthews 1995). Enhancement of murine T-cell blastogenesis by 5-HT has been shown to occur through 5-HT2 receptors (Young et al. 1993), whereas activation of human T-cells occurs through 5-HT1A receptors (Aune et al. 1994).

The effects of 5-HT on the production of cytokines by lymphocytes and macrophages was the subject of a few studies. Arzt et al. (1991) reported that 5-HT inhibits – in a concentration-dependent way – the LPS-induced production of TNF by human macrophages, and that this effect was blocked by the 5-HT2 receptor antagonist, ketanserin. The production of Th-1 cytokines, such as IL-2 and IFNγ, by Ag-stimulated, immune murine spleen cells is inhibited by 5HT1A receptor antagonists, but not by 5HT2 receptor antagonists (Aune et al. 1994). 5-HT effectively promotes IFNγ production in the presence of monocytes and 5-HT1A receptors transduce signals that are required for NK cells to produce IFNγ in response to IL-2 (Hellstrand et al. 1993; Hellstrand and Hermodsson 1993).

Effects of "Serotonergic" Antidepressants on Cytokine Production

Preliminary studies report that increased serum IL-1β, IL-6 and positive APP concentrations as well as increased mitogen-stimulated IL-2R and IFNγ production by peripheral blood mononuclear cells (PBMC) from depressed patients returned to normal in those patients who respond to chronic treatment with selective serotonin reuptake inhibitors (SSRIs), such as fluoxetine and sertraline, tricyclic antidepressants (TCAs) and/or lithium chloride (Sluzewska et al. 1995a,b; 1997b; Seidel et al. 1996; Maes et al. 1997a). On the other hand, in patients suffering from TRD, signs of IRS activation and lower plasma tryptophan are observed before as well as after prolonged treatment with SSRIs. These findings suggest a role of IRS activation in the pathophysiology of TRD (Sluzewska et al. 1995a, 1997b). Recovery of rats from depression-like states by prolonged administration of TCAs or SSRIs, e.g., in the olfactory bulbectomized or chronic mild stress models of depression, is accompanied by a normalization of IRS activation. For example, "clinical" recovery is associated with a normalization in serum APPs, such as albumin, and a normalization of the hyperproduction of IL-1 and IL-2 by splenocytes (Kubera et al. 1995, 1996; Song and Leonard 1994). Moreover, repeated administration (28 days) of imipramine, fluvoxamine and maprotiline increases the IL-1 receptor antagonist (IL-1RA) mRNA expression in a number of rat brain regions (Suzuki et al. 1996).

In vitro studies report that incubation of human PBMC or diluted whole blood with SSRIs (e.g., citalopram, sertaline), heterocyclic antidepressants, i.e., trazodone (a 5-HT2 antagonist and SRI) and TCAs, such as clomipramine and imipramine (which both have SRI capacities), produced a marked inhibition of IL-1β, IL-2, IL-6, TNFα, and IFNγ production and a significant increase in the production of IL-10, a negative immunoregulatory cytokine (Xia et al. 1996; Maes et al. 1999). Incubation of diluted whole blood with moclobemide, a reversible inhibitor of MAO_A (RIMA, which increases 5-HT in the synaptic cleft), venlafaxine, a serotonin-noradrenaline reuptake inhibitor (SNRI), and L-5-HTP, the precursor of 5-HT, increases the stimulated production of IL-10 (Lin et al. 1999; Kubera et al., submitted for publication). Thus, the results of the above studies suggest that various types of antidepressive drugs, which all modulate the serotonergic system, including SSRIs, SNRIs, TCAs, HCAs and L-5-HTP, may have negative immunoregulatory effects. A significant increase in IL-10 production is also observed in mice splenocytes after acute and chronic administration of desipramine (Kubera, 1999). In light of clinical and experimental findings, it is not unreasonable to suggest that antidepressants may alleviate depressive symptoms by their negative immunoregulatory effects (Connor and Leonard 1998).

Conclusions

In conclusion, major depression is accompanied by activation of the IRS and by disorders in the central and peripheral turnover of serotonin. There are significant relationships between signs of IRS activation and the lower availability of tryptophan to the brain, one of the most consistently reported abnormalities in depression. These findings suggest that the lowered availability of tryptophan to the brain may result from IRS activation, e.g., via induction of IDO by proinflammatory cytokines. Antidepressants with a serotonergic mode of action may normalize signs of IRS activation in depression. In vitro tests show that the antidepressants tested have immunoregulatory effects by increasing the production of interleukin-10 and by decreasing the interferon-γ/interleukin-10 production ratio. The mechanism by which antidepressants exert their negative immunoregulatory effects has remained elusive. It is suggested that the therapeutic efficacy of antidepressive drugs may, in part, be related to their negative immunoregulatory effects, which tend to normalize IRS activation in depression and, consequently, could lessen the load exerted by IRS activation upon the serotonergic system. These findings may be consistent with the hypothesis of a combined dysregulation of the serotonergic system and IRS activation as an integral component of major depression.

Acknowledgments. The reported research was supported by the Flemish Community, Belgium; the Funds for Scientific Research, Vlaanderen, Belgium (FWO); the Clinical Research Center for Mental Health, (CRC-MH), Antwerp, Belgium; the Staglin Investigator Award (NARSAD, USA) to Dr. M. Maes, and grant no. PO5A098-12 from the Committee on Scientific Research, Poland. The assistance of E. Nowak, Mrs. M. Maes and R. Bullock are acknowledged.

References

Arzt E, Costas M, Finkielman S, Nahmod VE (1991) Serotonin inhibition of tumor necrosis factor-alpha synthesis by human monocytes. Life Sci 48:2557–2562

Aune TM, Golden HW, McGrath KM (1994) Inhibitors of serotonin synthesis and antagonists of serotonin 1A receptors inhibit T lymphocyte function in vitro and cell-mediated immunity in vivo. J Immunol 153:489–498

Barkai AI, Baron M, Kowalik S, Cooper TB (1986a) Modification of ligand binding to membranes by a soluble acceptor. Alpha-1-acid glycoprotein attenuates 3H-imipramine binding to cerebral membranes. Biol Psychiat 21:883–888

Barkai AI, Baron M, Kowalik S, Cooper TB (1986b) Inhibition of 3H-imipramine binding by plasma from depressed and normal subjects. Psychiatr Res 17:261–267

Bonaccorso S, Lin A, Song C, Verkerk R, Kenis G, Bosmans E, Scharpe S, Vandewoude M, Dossche A, Maes M (1998) Serotonin-immune interactions in elderly volunteers and in patients with Alzheimer's disease (DAT): lower plasma availability of tryptophan in elderly subjects and increased serum interleukin-6 in DAT. Aging Clin Exp Res 10:316–323

Bonnet M, Lespinats C, Burtin C (1984) Histamine and serotonin suppression of lymphocyte response to phytohemagglutinin and allogeneic cells. Cell Immunol 83:280–291

Clement HW, Buschmann J, Rex S, Grote C, Opper C, Gemsa D, Wesemann W (1997) Effects of interferon-gamma, interleukin-1beta, and tumour necrosis factor-alfa on the serotonin metabolism in the nucleus raphe dorsalis of the rat. J Neural Transm 104:981–991

Connor TJ, Leonard BE (1998) Depression, stress and immunological activation: the role of cytokines in depressive disorders. Life Sci 62:583–606

Devoino LV, Morozova NB (1988) Serotoninergic system in neuroimmunomodulation. Mechanism of inhibition. In: Novera H (ed) Neuroimmunomodulation. Proceedings of the first international workshop. Spector, Gordon and Breach Science Publishers, New York 119–123

Devoino L, Korovina L, Ilyutchenok R (1968) Influence of some drugs on the immune response. I. The effect of 5-hydroxytryptophan, iproniazid and reserpine on primary and secondary immune responses. Eur J Pharmacol 4:441–448

Dunn AJ (1992) Endotoxin-induced activation of cerebral catecholamine and serotonin metabolism: comparison with interleukin-1. J Pharmacol Exp Ther 261:964–969

Hellstrand K, Hermodsson S (1993) Serotonergic 5HT1A receptors regulate a cell contact-mediated interaction between natural killer cells and monocytes. Scand J Immunol 37:7–18

Hellstrand K, Czerkinsky C, Ricksten A, Jansson B, Asea A, Kylefjord H, Hermodsson S (1993) Role of serotonin in the regulation of interferon-gamma production by human natural killer cells. J Interferon Res 13:33–38

Heyes MP, Saito K, Crowley JS, Davis LE, Demitrack MA, Der M, Dilling LA, Elia J, Kruesi MJP, Lackner A, Larsen SA, Lee K, Leonard HL, Markey SP, Martin A, Milstein S, Mouradian MM, Pranzatelli MR, Quearry BJ, Salazar A, Smith M, Strauss SE, Sunderland T, SW, Swedo Tourtellotte WW (1992) Quinolinic acid and kynurenine pathway metabolism in inflammatory and non-inflammatory neurological disease. Brain 115:1249–1273

Irwin MR, Smith TL, Gillian JC (1987) Reduced natural killer cytotoxicity and depressed patients. Life Sci 41:2127–2133

Kouhata S, Kagaya A, Nakae S, Yokota N, Horiguchi J, Nakata Y, Yamawaki S (1998) Effect of acute administration of lipopolysaccharide on DOI-induced wet dog shaking behavior in rats. Abstract PT08040 XXIst CINP Congress, Glasgow, 1998.

Kronfol Z, Silva J, Greden J, Dembinski S, Carroll B (1983) Impaired lymphocyte function in depressive illness. Life Sci 33:241–247

Kubera M, Basta-Kaim A, Papp M (1995) The effect of chronic treatment with imipramine on the immunoreactivity of animals subjected to a chronic mild stress model of depression. Immunopharmacology 30:225–230

Kubera M, Simbirtsev A, Basta-Kaim A, Borycz J, Roman A, Papp M, Claesson M (1996) Effect of chronic treatment with imipramine on IL-1 and IL-2 production by splenocytes obtained from rats subjected to a chronic mild stress model of depression. Pol J Pharmacol 48:503–506

Lin A, Song C, Kenis G, Bosmans E, De Jongh R, Scharpe S, Maes M (1999) The in vitro immunosuppressive effects of moclobemide. J Affect Disord, in press

Maes M (1993) Acute phase protein alternations in major depression: a review. Rev Neurosci 4:407–416

Maes M (1997) The immune pathophysiology of major depression. In: Honig A, van Praag HM (eds) Depression: neurobiological, psychopathological and therapeutic advances. John Wiley, London, 197–215

Maes M, Meltzer HYM (1995) The serotonin hypothesis of major depression. In: Bloom F, Kupher D (eds) Psychopharmacology: the fourth generation of progress. Raven Press, New York, 933–944

Maes M, Meltzer HY, Scharpe S, Bosmans E, Suy E, Minner B, De Meester I, Calabrese J, Vandervorst C, Raus J, Cosyns P (1993) Relationships between lower plasma L-tryptophan levels and immune variables in depression. Psychiatr Res 49:11–27

Maes M, Smith R, Scharpe S (1995) The monocyte-T-lymphocyte hypothesis of major depression. Psychoneuroendocrinology 20:111–116

Maes M, Wauters A, Verkerk R, Neels H, vanGastel A, Cosyns P, Scharpe S, Desnyder R (1996) Lower L-tryptophan availability in depression: a marker of a more generalized disorder in protein metabolism. Neuropsychopharmacology 15:243–251

Maes M, Delanghe J, Ranjan R (1997a) Acute phase proteins in schizophrenia, mania and major depression: modulation by psychotropic drugs. Psychiatr Res 66:1–11

Maes M, Verkerk R, Vandoolaeghe E, Van Hunsel F, Neels H, Wauters A, Demedts P, Scharpe S (1997b) Serotonin-immune interactions in major depression: lower serum tryptophan as a marker of an immune-inflammatory response. Eur Arch Psychiat Clin Neurosci 247:154–161

Maes M, Lin A, Bosmans E, Vandoolaeghe E, Bonnacorso S, Kenis G, De Jongh R, Verkerk R, Scharpe S (1998) Serotonin-immune interactions in detoxified chronic alcoholic patients without apparent liver disease: indicants of immune activation and lower availability of plasma total tryptophan. Psychiatr Res 78:151–161

Maes M, Song C, Lin A, Bonaccorso S, Scharpe S, Kenis G, Bosmans E (1999) Antiinflammatory effects of antidepressants (fluoxetine, trazodone, clomipramine): increased production of interleukin-10 and suppressed production of interferon-gamma. Neuropsychopharmacology, in press

Maier SF, Watkins LR (1998) Cytokines for psychologists: implications of bidirectional immune-to-brain communication for understanding behaviour, mood, and cognition. Psychol Rev 105: 83–107

Nemeroff CB, Krishnan KR, Blazer DG, Knight DL, Benjamin D, Meyerson LR (1990) Elevated plasma concentrations of alpha 1-acid glycoprotein, a putative endogenous inhibitor of the tritiated imipramine binding site, in depressed patients. Arch Gen Psychiat 47:337–340

Ptak W, Herzog WR, Askenase PW (1991) Delayed-type hypersensitivity initiation by early-acting cells that are antigen mismatched or MHC incompatible with late-acting, delayed-type hypersensitivity effector T cells. J Immunol 146:469–475

Ramamoorthy S, Ramamoorthy JD, Prasad PD, Bhat GK, Mahesh VB, Leibach FH, Ganapathy V (1995) Regulation of the human serotonin transporter by interleukin-1 beta. Biochem Biophys Res Commun 216:560–567

Seidel A, Arolt V, Hunstiger M, Rink L, Behnisch A, Krichner H (1996) Cytokine production and serum proteins in depression. Scand J Immunol 41:534–538

Sluzewska A, Rybakowski JK, Laciak M, Mackiewicz A, Sobieska M, Wiktorowicz K (1995a) Interleukin-6 levels in depressed patients before and after treatment with fluoxetine. Ann NY Acad Sci 762:474–477

Sluzewska A, Rybakowski JK, Sobieska M, Bosmans E, Pollet H, Wiktorowicz K (1995b) Increased levels of alpha-1-acid glycoprotein and interleukin-6 in refractory depression. Depression 3:170–175

Sluzewska A, Samborski W, Sobieska M, Klein R, Bosmans E, Rybakowski JR (1997a) Serotonin antibodies in relation to immune activation in major depression. Psychiatr Res 12:453–458

Sluzewska A, Sobieska M, Rybakowski JK (1997b) Changes in acute-phase proteins during lithium potentation of antidepressants in refractory depression. Neuropsychobiology 35:123–127

Smith RS (1991) The macrophage theory of depression. Med Hypoth 35:298–306

Song C, Leonard BE (1994) Serotonin reuptake inhibitors reverse the impairments in behaviour, neurotransmitter and immune functions in the olfactory bulbectomized rat. Human Psychopharmacol 9:135–146

Strijewski A, Chudzik J, Tang SW (1988) Inhibition of platelet [3H]-imipramine binding by human plasma protein fractions. Life Sci 42:1543–1550

Suzuki E, Shintani F, Kanba S, Asai M, Nakaki T (1996) Induction of interleukin-1 beta and interleukin-1 receptor antagonist mRNA by chronic treatment with various psychotropics in widespread areas of rat brain. Neurosci Lett 215:201–204

Xia Z, DePierre JW, Nassberger L (1996) Tricyclic antidepressants inhibit IL-6, IL-1-beta and TNF-alfa release in human blood monocytes and IL-2 and interferon-gamma in T cells. Immunopharmacology 34:27 – 37

Young MRI, Matthews JP (1995) Serotonin regulation of T-cell subpopulations and of macrophage accessory function. Immunology 84:148-152

Young MRI, Kut JL, Coogan MP, Wright MA, Young ME, Matthews J (1993) Stimulation of splenic T-lymphocyte function by endogenous serotonin and by low-dose exogenous serotonin. Immunology 80:395–400

Responses in the Raphe-Hippocampal Serotonergic System During Peripheral Inflammation: Putative Role in Sickness Behaviour

J. M. H. M. Reul and A. C. E. Linthorst

Summary

It is now well established that challenges to the immune system trigger a number of physiological processes, such as activation of the hypothalamic-pituitary-adrenocortical (HPA) axis, fever, and sickness behaviour. The task of these triggered processes is to promote survival and recovery of homeostasis. However, at the level of the brain the neurotransmitter systems underlying these physiological and behavioural responses still need to be fully characterized. By use of an in vivo microdialysis method in rats, we have found that during peripheral inflammation (induced by intraperitoneal (ip) injection of endotoxin (lipopolysaccharide; LPS) marked changes occur in serotonergic neurotransmission in the brain. Hence, a profound increase in extracellular levels of serotonin (5-HT) was observed in the hippocampus after an ip injection of LPS. This endotoxin effect was mainly mediated by interleukin-1 (IL-1). Based on several studies, it was postulated that the rise in hippocampal 5-HT levels might be associated with the development of sickness behaviour. Moreover, a peripheral inflammatory challenge was shown to evoke a long-term increase in 5-HT_{1A} receptor density in hippocampal subregions but not in the raphe area.

Disturbances in the functioning of the immune system described in major depression prompted us to study whether disturbed interactions between the central nervous system and the immune system may be implicated in the development of stress-related disorders such as major depression. The observation that depressed patients show an elevated central drive of corticotropin-releasing hormone (CRH) in conjunction with the fact that this neuropeptide is highly involved in physiological responses to stress lead us to develop an animal model of CRH hyperactivity, i. e., the long-term, intracerebroventricularly (icv) CRH-infused rat. In our experimental design, rats were challenged ip with LPS after seven days of icv CRH or vehicle infusion. The endotoxin evoked a profound fever response in control animals, whereas only a blunted response developed in long-term CRH-treated rats. Moreover, LPS produced an attenuated hippocampal 5-HT response and a delayed onset of HPA axis activation and behavioural inhibition in CRH-treated rats. These attenuated, brain-mediated responses could not have been caused by suppression of the release of plasma cytokines, since CRH-infused animals showed enhanced LPS-induced plasma bioactivities of IL-1 and IL-6 (but not of tumor necrosis factor (TNF)).

Patterson/Kordon/Christen (Eds.)
Neuro-Immune Interactions in Neurologic and Psychiatric Disorders

These studies indicate that peripheral inflammation elicits vast changes in hippocampal serotonergic neurotransmission, both in terms of serotonin release and 5-HT_{1A} receptor density, that might be responsible for the initiation of specific physiological, behavioural and molecular responses in order to cope with the challenge. In addition, chronically elevated CRH drive in the brain, as seen during chronic stress and stress-related disorders results in aberrant brain-mediated responses to an acute inflammatory challenge – a situation which, in the long run, may present a detrimental threat for health. Thus, these results provide evidence that chronic CRH hypersecretion is an important factor in the etiology of stress-related disorders.

Introduction

During the last two decades, research on the immune system and the nervous system has overturned two dogmas. One dogma stated that the brain is an immune-privileged organ; the other stated that the immune system is an autonomously operating entity functioning without external influences. It is now clear that these two systems interact intensely with each other. Looking back, the evidence for the existence of such interactions has always been present in our every day life, given the psychological, behavioural and physiological manifestations we note during times of illness.

Many aspects of the wide variety of defence mechanisms with which an organism responds to infectious or inflammatory agents have been delineated. As a first step, the immune system is activated, resulting in, among other things, the production of various cytokines. Subsequently, some of these cytokines are able to raise the levels of circulating glucocorticoids, which accelerate the course of the immune response and prevent it from overshooting (Munck et al. 1984; Wiegers and Reul 1998; Wiegers et al. 1995). Cytokines released during moderate to severe infectious and inflammatory states produce an array of physiological and behavioural responses such as fever, tachycardia, loss of appetite, behavioural inhibition, fatigue and social disinterest. The behavioural changes are collectively termed "sickness behaviour" (Dantzer et al. 1991; Hart 1988), which is thought to help overcome the infection or inflammation and thereby to increase the survival chances.

Although the physiological, endocrine and behavioural responses to immune challenges have been extensively described during the last years, the neurochemical mechanisms in the brain underlying these phenomena still need to be clarified. Therefore, several years ago we embarked on characterizing how the central neurotransmitter systems respond to peripheral inflammatory stimulation. Moreover, our aim was to elucidate the functional role of changes in specific neurotransmitter activities during inflammatory challenges. Our interest in studying neuroimmune interactions was actually twofold. First, the prime focus of our research concerns the study of how the brain processes stress and, in this respect, inflammation or immune stimulation in general can be regarded as a stressful event. Second, from a psychiatric perspective, immune system changes have been observed in patients suffering from major depression (Irwin et al. 1992; Maes 1995; Sluzewska et al. 1996). Furthermore, one of the typical characteristics of major depression is an increased central drive of corticotropin-releasing hormone (CRH; for references see Holsboer and Barden 1996). CRH is a neuropeptide that plays a principal role in the brain as a mediator of neuroendocrine, autonomic and behavioural responses to stressful stimuli (for references see Vale et al. 1997). It is also implicated in the regulation of immune system functioning (Irwin et al. 1987). Because we hypothesized that aberrant neuroimmune interactions may be implicated in the precipitation of psychiatric diseases such as major depression in subjects with chronic central CRH hypersecretion (occurring congenitally and/or during chronic stress), we studied these interactions in a rat model with long-term elevated levels of CRH in the brain.

In this chapter, we describe the effects of peripheral inflammation on brain neurotransmitter systems with the main focus on serotonin (5-HT). The role of specific cytokines in inflammation-induced neurotransmitter changes will be discussed on the basis of the results of our own experiments as well as of data provided by other research groups. Finally, the implications of a long-term increased drive of CRH for immune system functioning, brain-mediated physiological and behavioural responses, and underlying neurochemical changes are described.

Integrative Approach Through Parallel Collection of Neuropharmacological, Neuroendocrine and Behavioural Data

Until recently, the neurochemical basis of the physiological and behavioural responses evoked by peripheral inflammatory stimuli had hardly been studied. Previously, in most studies concentrations of neurotransmitters and their metabolites were determined in extracts of brain tissue samples and calculated as an index of turnover. Unfortunately, such studies provide only limited information about the extracellular (i.e., biologically effective) levels of neurotransmitters and the time resolution of changes. Therefore, we applied an in vivo microdialysis technique to rats, which allows the simultaneous assessment of different neurotransmitters and metabolites and the continuous monitoring of neurotransmitters over extended time periods (hours and days).

Furthermore, a method was developed to determine corticosterone levels in dialysates, providing an index of the net hypothalamic-pituitary-adrenocortical (HPA) axis output. The dialysate concentrations of corticosterone indeed represent the free (biologically active) fraction of this glucocorticoid because the extracellular space of the brain is devoid of the transport protein corticosterone binding globulin. For the correct interpretation of the serotonin data (see below) and the assessment of sickness behaviour, the behavioural activity (locomotion, rearing, grooming, eating, drinking, resting) of the rats was monitored by visual observation of the animals.

Our interest in measuring putative changes in neurotransmitter release in higher limbic brain structures was triggered because inflammatory insults and infections, apart from immune, neuroendocrine and autonomic responses, are also known to induce changes in the psychological state (i.e., vigilance state, mood and affect) of subjects. Thus, we implanted the microdialysis probes primarily in the hippocampus, a limbic brain region that plays a critical role in the cognitive interpretation of (stressful) sensory information. As a model for peripheral inflammation, rats were injected intraperitoneally (ip) with lipopolysaccharide (LPS), which is a cell wall constituent of gram-negative bacteria. Moreover, to delineate the role of specific cytokines in these effects, certain cytokines or cytokine receptor antagonists were applied. Data from our laboratory and others will be reviewed and discussed.

Peripheral Inflammation and Hippocampal Serotonergic Neurotransmission

Induction of a peripheral inflammatory response by ip injection of LPS (30-300 μg/kg body weight) evoked a dose-dependent increase in hippocampal extracellular levels of 5-HT (maximum between 200–250% of baseline) and its metabolite 5-hydroxyindoleacetic acid (5-HIAA; maximum about 150% of baseline; Linthorst et al.1995b, 1996). The doses of LPS used in this and our subsequent studies (applying 100 μg/kg body weight) induced fever responses and marked reductions in behavioural activity of the animals, in addition to other signs of sickness behaviour, such as a curled-up body posture and piloerection. LPS also stimulated the activity of the HPA axis as indicated by a vast and extended rise in free corticosterone levels in the extracellular fluid of the brain (Linthorst et al. 1995a, b).

The extracellular levels of 5-HIAA rose considerably as a result of the peripheral inflammatory stimulus (Linthorst et al. 1995b). 5-HIAA levels in the extracellular fluid reflect the amount of newly synthesized 5-HT in the synaptic terminals and do not represent an index of the metabolism of released 5-HT (Auerbach et al. 1989; Crespi et al. 1990; Kalén et al. 1988; Linthorst et al. 1994). Thus, the increases in 5-HIAA levels indicate that peripheral inflammatory insults result in an enhanced synthesis of this indoleamine.

Merali and colleagues (1997) have shown in recent microdialysis experiments that ip injected IL-1 also produced elevations in extracellular levels of 5-HT and 5-HIAA in the hippocampus. Earlier studies on neurotransmitter levels in brain tissue extracts of rats and mice demonstrated that systemic application of IL-1 (Kabiersch et al. 1988; Zalcman et al. 1994) and IL-6 (Zalcman et al. 1994) increased the turnover of 5-HT (as indicated by a rise in the ratio of 5-HIAA/5-HT) and/or the levels of 5-HIAA. Together these findings indicate that, following ip inoculation of LPS, the peripherally produced IL-1 (and IL-6) can affect serotonergic neurotransmission in the hippocampus.

The effects of peripheral inflammation on higher brain structures are not confined to the hippocampus. Studies on neurotransmitter levels in brain tissue extracts have shown that peripheral injection of endotoxin, IL-1 and IL-6 enhanced the turnover of 5-HT in the frontal cortex (Dunn 1992; Dunn and Welch 1991; Zalcman et al. 1994). Moreover, increased levels of 5-HIAA were found in dialysates from the frontal cortex after ip administration of LPS (Lavicky and Dunn 1995). Unfortunately, no information on the parent amine (5-HT) was provided in this study.

In many actions of cytokines, arachidonic acid metabolites play a major (i.e., intermediary) role (Gottschall et al. 1992; Rothwell 1991). Systemic administration of LPS has been shown to induce the release of prostaglandin E_2 (PGE_2) in the hypothalamus (Sirko et al. 1989; Smith et al. 1994). Moreover, IL-1β raised the levels of PGE_2 in rat astrocyte cultures and rat brain (Katsuura et al. 1989; Komaki et al. 1992). The cyclo-oxygenase inhibitor indomethacin (i.e., an inhibitor of prostaglandin production) restrained the IL-1-induced release of adrenocorticotropin (ACTH; Katsuura et al. 1988). Prostaglandins also participate in changes in neurotransmission during inflammation. Pretreatment of rats with indomethacin (10 mg/kg body weight ip) almost completely abolished the LPS-induced increase in hippocampal serotonergic neurotransmission (Linthorst et al. 1996). The prostaglandin synthesis inhibitor, however, succeeded in entirely preventing the fever, tachycardia and sickness behaviour responses. Thus, the partial inhibition of the LPS-evoked rise in hippocampal 5-HT induced by indomethacin suggests that other mediators may be involved. Nitric oxide might be a candidate, because results of various studies point a role for this gaseous molecule in central nervous responses to inflammation (Karanth et al. 1993; Raber and Bloom 1994; Rivier and Shen, 1994).

Brain Cytokines and Hippocampal Serotonergic Neurotransmission

From the previous sections it is clear that, during a peripheral inflammatory response, an activation of serotonergic neurotransmission in the hippocampus and, possibly, other parts of the brain occurs. Because the synthesis and release of many (proinflammatory) cytokines are induced during peripheral inflammation we needed to clarify which cytokine(s) were involved in each LPS-evoked response. Therefore, a series of studies was designed to delineate which cytokines participated in the altered hippocampal serotonergic neurotransmission seen during peripheral inflammation. First, we tested whether single icv injections of selected cytokines were able to mimic the effects of ip-injected LPS. Two proinflammatory cytokines were examined – tumor necrosis factor-α (TNF-α) and IL-1β – and their action was compared with that the non-inflammatory cytokine IL-2. Biotelemetry measurements showed that icv administration of all three cytokines increased body temperature and free corticosterone levels in the dialysates (Kluger 1991; Linthorst et al. 1995b; Pauli et al. 1998). However, whereas IL-1β and IL-2 evoked sickness behaviour and profound increases in hippocampal extracellular levels of 5-HT and 5-HIAA, TNF-α had no effects on these parameters (using doses effective in inducing fever and HPA axis activation; Linthorst et al. 1995b; Pauli et al. 1998). A recent study on neurotransmitter levels in brain tissue extracts showed that icv administration of IL-1β increased hippocampal 5-HIAA levels in rats, whereas IL-2, IL-6 and TNF-α were ineffective (Connor et al. 1998). The seeming inconsistency between the latter publication and our observation regarding the effects of IL-2 may be explained by methodological differences (brain tissue levels vs. in vivo microdialysis) and also by the earlier time point of killing of the rats (45 min after administration of IL-2) in the study by Connor et al. (1998). Thus, it seems that both central IL-1β and IL-2, but not TNF-α, are able to stimulate hippocampal serotonergic neurotransmission.

IL-2 is an important cytokine within the immune system, where it acts as the main T-cell growth factor. The cytokine also stimulates the release of various cytokines from immune cells, including IL-1 (Janssen et al. 1994). As the effects of IL-2 on body temperature, HPA axis activity, behaviour and hippocampal serotonin resembled those of IL-1, we contemplated that IL-2 might exert at least some of its effects on the level of the brain via the induction of IL-1. Therefore, rats were pretreated icv with the IL-1 receptor antagonist (IL-1ra) prior to icv administration of IL-2. IL-1ra completely abrogated the IL-2-induced increases in hippocampal 5-HT and 5-HIAA, but only attenuated the increase in free corticosterone (Pauli et al. 1998). Thus, the effects of IL-2 on hippocampal serotonergic neurotransmission are mediated by centrally released IL-1, whereas this proinflammatory cytokine seems to be only partially involved in the effects of IL-2 on the HPA axis.

The early observation that IL-1β is one of the principal LPS-induced proinflammatory cytokines and our own finding that central injection of IL-1β exerts pronounced effects on hippocampal 5-HT prompted us to investigate whether brain IL-1 might be a mediator of the LPS-evoked changes in the levels of this neurotransmitter. Indeed, when rats were pretreated icv with IL-1ra (10 μg), we observed a 50% reduction of the ip LPS-induced activation of hippocampal 5-HT release, strongly suggesting that centrally released IL-1 is a prime mediator of the effects of LPS on 5-HT in the hippocampus (Linthorst et al. 1995b). Moreover, local infusion of IL-1β into the hippocampus (apart from HPA axis activation and induction of sickness behaviour) also resulted in marked elevations of extracellular 5-HT levels (Linthorst et al. 1994). These data support the notion that IL-1β released/produced locally in the hippocampus participates in the activation of 5-HT release in this brain structure during peripheral inflammation.

Peripheral Inflammation and Raphe-Hippocampal 5-HT$_{1A}$ Receptors

One of the most abundant serotonin receptor subtypes in the hippocampus is the 5-HT$_{1A}$ receptor. In this brain structure, the 5-HT$_{1A}$ receptor is located post-synaptically whereas in the raphe nuclei, i.e., the serotonergic cell body regions, this receptor subtype functions as a somatodendritic autoreceptor. We examined whether the marked activation of hippocampal serotonergic neurotransmission following peripheral inflammation would have any consequences for the 5-HT$_{1A}$ receptor density in the subregions of the hippocampus and the median and dorsal raphe nuclei. In the median and dorsal raphe nucleus as well as in some subregions of the hippocampus, the 5-HT$_{1A}$ receptor densities displayed a circadian variation (Fig. 1). This phenomenon has also been indicated by experiments of other investigators (Nagayama and Lu, 1997). However, we show here for the first time, that LPS evoked increases in the density of this receptor subtype in various subregions of the hippocampus which were revealed at least 12 to 24 hours post-injection (Fig. 1). In contrast, the raphe nuclei did not show any changes in 5-HT$_{1A}$ receptor densities following the LPS injection (Fig. 2). The receptor density data obtained in the hippocampus were rather unexpected in view of the profound release of 5-HT seen after LPS administration (Linthorst et al. 1995b). Therefore, these results clearly indicate that, during peripheral inflammation, the hippocampal 5-HT$_{1A}$ receptors are not subject to homologous downregulation, but are even upregulated by an as yet unknown mechanism. Obviously, such a mechanism is not operational in the raphe nuclei. Taken together, these observations suggest that, after an inflammatory challenge, the rise in extracellular concentrations of 5-HT together with the stimulation of its synthesis (as indicated by 5-HIAA) and the upregulation of 5-HT$_{1A}$ receptors may warrant a profound and extended activation of 5-HT-regulated postsynaptic events in the hippocampus.

Site-Specific Neuronal Activation Within the Raphe-Hippocampal System During Peripheral Inflammation: Evidence from c-fos Studies and Functional Implications

Various research groups have reported the expression of c-fos in the brain after peripheral administration of endotoxin or IL-1 (Brady et al. 1994; Elmquist et al. 1993; Ericsson et al. 1994; Sagar et al. 1995; Wan et al. 1993; A. Bilang-Bleuel, A.C.E. Linthorst, and J.M.H.M. Reul, unpublished observations). To our knowledge, until now no study has reported on the existence of c-fos expression in the raphe nuclei after inflammatory stimulation. There are two additional findings that also indicate that the raphe nuclei do not respond to immune system challenges. Recently, Molina-Holgado and Guaza (1996) reported that ip administration of LPS had no effect on the tissue levels of NA and 5-HT in the raphe nuclei, but the levels of both neurotransmitters were found to be increased in the nucleus of the solitary tract (NTS).

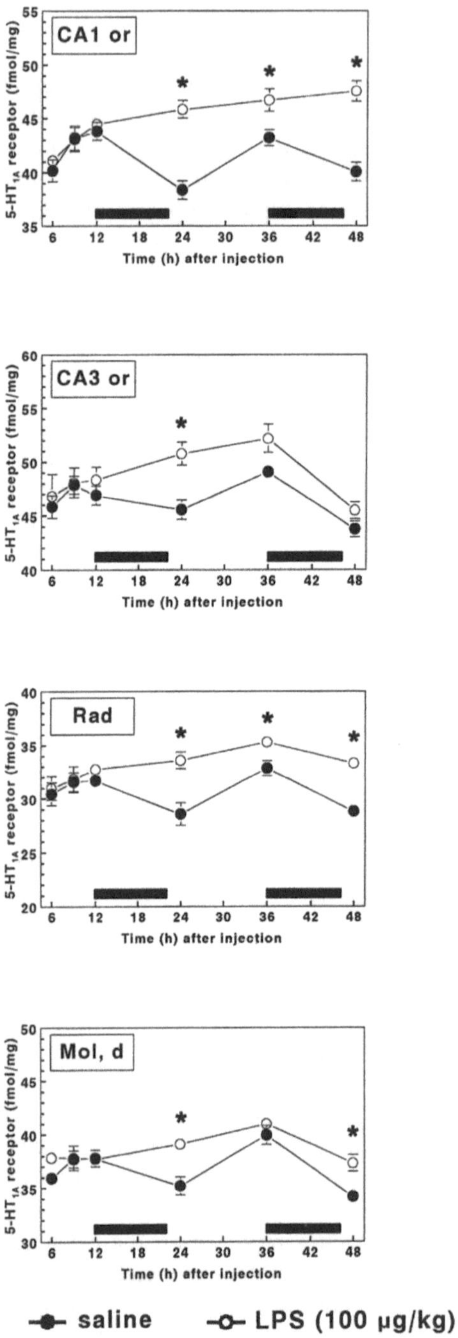
CA1 or
CA1 py
CA3 or
CA3 py
Rad
Gr DG
Mol, d
Mol, v
5-HT$_{1A}$ receptor (fmol/mg)
Time (h) after injection
saline
LPS (100 µg/kg)

Moreover, as previously described, we have shown that ip administration of LPS increased 5-HT_{1A} receptor density in various regions of the hippocampus, but did not affect the autoreceptor levels in both the dorsal and median raphe nucleus. However, although LPS at a dose of 100 μg/kg body weight had no effect on extracellular levels of 5-HT in the preoptic area (Linthorst et al. 1995a), this dose produced a marked increase in the hippocampus. Here the effects on 5-HT release appear to be instigated by a local action of IL-1 (Linthorst et al. 1994, 1995b). Thus, it seems that IL-1-involved mechanisms at the *terminal* level in the hippocampus are instrumental in the LPS-evoked increases in 5-HT in this brain structure.

One of our principal aims was to elucidate the functional implications of the hippocampal serotonergic system activation during peripheral inflammation. We have postulated that the rise in hippocampal 5-HT levels might be instrumental for the establishment of (some aspects of) sickness behaviour (Reul et al. 1998). Among other functions, serotonin is known to modulate stress-related behaviour and to regulate the processing of sensory information. Several investigations in cats and rats have shown that a positive correlation exists between the animal's behavioural activity and the firing rate of serotonergic neurons and extracellular levels of 5-HT in the terminal areas (Jacobs and Azmitia 1992; Kalén et al. 1989; Linthorst et al. 1994). The strongest correlation between extracellular 5-HT and behavioural activity was observed in the hippocampus, although such a relationship was also evident in the preoptic area and neocortex (Linthorst et al. 1995a,b). Various stressors increase hippocampal extracellular levels of 5-HT in parallel with behavioural activation (Jacobs and Azmitia 1992; Kalén et al. 1989; Rueter and Jacobs 1996). However, the situation during inflammatory stress is unique, because here a *rise* in hippocampal 5-HT levels was seen during a period of *decreased* behavioural activity (sickness behaviour). As pointed out previously, this effect seems to be generated solely at the level of the hippocampus; i.e., without a concomitant increase in the firing rate of the serotonergic cell bodies.

Fig. 1. Effect of ip LPS on 5-HT_{1A} receptor densities in the subregions of the rat hippocampus as revealed by in vitro autoradiography. Rats were injected ip with LPS (type *Salmonella abortus equi*; 100 μg/kg) and killed after various time points up to 48 hours post-injection. Brains were collected for preparation of cryosections (20 μm thickness) and in vitro autoradiography was performed to visualize and quantify the 5-HT_{1A} receptors. The receptors were labelled for determination of the total binding with 0.07 nM of the 5-HT_{1A} agonist [^{125}I]-8-hydroxy-2-[N-n-propyl-N-(3′-iodo-2′-propylen)-amino] tetralin ([^{125}I]-8-OH-PIPAT) in combination with 0.5 μM Ritanserin to block 5-HT_7 and 5-HT_{2A} receptors. Nonspecific binding was determined by inclusion of 10 μM ±8-OH-DPAT in the incubation solution. Receptor densities were assessed by semi-quantitative image analysis and expressed as fmol/mg protein (mean ± standard error of the mean, n = 3−6). The black bars indicate the dark periods of the light-dark cycle. Data were statistically analysed by two-way analysis of variance and post-hoc Duncan multiple range test. *, $P < 0.05$. Abbreviations: CA1−3, cellular fields of Ammon's horn; DG, dentate gyrus; GrDG, granular layer of the dentate gyrus; CA1 or, oriens layer of the CA1; CA1 py, pyramidal layer of the CA1; CA3 or, oriens layer of the CA3; CA3 py, pyramidal layer of the CA3; Rad, stratum radiatum; Mol, d, dorsal molecular layer of the dentate gyrus; Mol, v, ventral molecular layer of the dentate gyrus

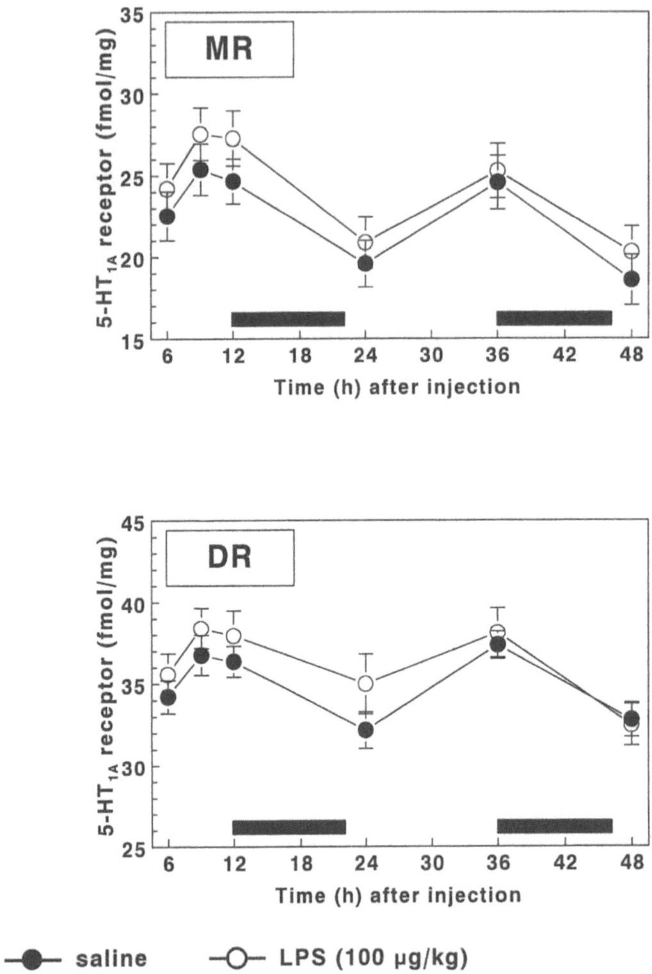

Fig. 2. Effect of ip LPS on 5-HT_{1A} receptor densities in the median and dorsal raphe nucleus as revealed by in vitro autoradiography. For experimental details, see legend to Figure 1. Data are expressed as fmol/mg protein (mean ± standard error of the mean, n = 3–6). Data were statistically analysed by two-way analysis of variance and post-hoc Duncan multiple range test. *, $P < 0.05$. Abbreviations: MR, median raphe nucleus; DR, dorsal raphe nucleus

We have proposed that, in this manner, a situation is realized in which sensory information processing in the hippocampus is modulated by 5-HT, but a parallel increase in behavioural activity by facilitation of motor output by 5-HT at the level of the brain stem is prevented (Reul et al. 1998). An involvement of hippocampal 5-HT in sickness behaviour is underscored by the observation that behavioural lethargy after an inflammatory challenge is always associated with elevated hippocampal levels of 5-HT (for instance in the case of LPS, IL-1β, IL-2). TNF-α at doses that cause HPA axis activation and body temperature elevation, however, neither generates a reduction in behavioural activity nor raises hippocampal 5-HT concentrations (Pauli et al. 1998). Additional support for this hypothesis stems from neuroimmune experiments on long-term icv CRH-infused rats. These rats show an attenuated rise in hippocampal 5-HT levels during peripheral inflammation combined with a delayed onset of sickness behaviour (see below).

These observations indicate that hippocampal 5-HT is implicated in sickness behavior, but it should be noted that this concept is mainly based on correlative evidence. Future experiments applying specific 5-HT receptor antagonists, antisense oligodeoxynucleotide treatments, lesion paradigms and animals with selected mutations are required to further define the role of brain 5-HT in sickness behavior.

Neuroimmunoendocrine Interactions in Rats Chronically Exposed to Elevated Levels of CRH in the Brain

CRH is a neuropeptide that is widely distributed in the central nervous system and functions as the principal mediator of neuroendocrine (mainly affecting the HPA axis), autonomic (e.g., sympathetic outflow) and behavioural responses of the organism to stress (for references see Vale et al. 1997). Furthermore, icv injection of CRH leads to a reduction in natural cell activity, which indicates that CRH also plays a role in immune regulation (Irwin et al. 1987).

Evidence has accumulated that malfunctioning of brain CRH system(s) plays, an important role in stress-related disorders such as major depression and anorexia/bulimia, but also in Alzheimer's disease. Hence, it appears that major depression is associated with an increased central CRH drive. Patients suffering from major depression show increased levels of CRH in cerebrospinal fluid (Nemeroff et al. 1984) and increased numbers of CRH-expressing neurons and CRH mRNA levels in their paraventricular nucleus (PVN) (Raadsheer et al. 1994, 1995). Moreover, a decreased density of CRH receptor binding has been found in the frontal cortex of suicide victims (Owens and Nemeroff 1993). Importantly, depressed patients show many symptoms that are similar to the effects of centrally administered CRH in experimental animals, such as an activated HPA axis, increased sympathetic outflow, anxiety, increased emotionality, and loss of appetite and sexual interest (Owens and Nemeroff 1991; Holsboer and Barden 1996). It is often thought that the CRH hyperactivity seen in depressed patients is the consequence of a diminished glucocorticoid feedback efficacy via central glucocorticoid receptors. That this is not necessarily the case was reported in a recent publication of Dijkstra et al. (1998). We found that transgenic mice with impaired glucocorticoid receptor function did not show an increased expression of CRH but, in contrast, a marked reduction in the levels of this neuropeptide in the paraventricular nucleus of the hypothalamus and external zone of the median eminence (Dijkstra et al. 1998).

Based on observations in the cerebral cortex of deceased Alzheimer patients (decreased CRH content and increased CRH receptor binding), a reduced CRH activity has been proposed in the pathophysiology of Alzheimer's disease (De Souza et al. 1986). Moreover, the degree of cognitive impediment in Alzheimer patients appears to correlate with the extent of reduction in the cerebrospinal fluid CRH content (Pomara et et al. 1989).

With regard to the distinct functions of CRH and its involvement in certain psychiatric and neurological diseases, it seems that a well-balanced CRH system is vital for health. Nevertheless, although a disturbed functioning of CRH has been proposed in major depression and Alzheimer's disease, until now studies in humans have not been able to identify which of the physiological and behavioural abnormalities, associated with these illnesses are caused by hyper- or hyposecretion of CRH. Experimental paradigms for the investigation of central CRH dysfunction in humans are rather limited. Despite the extensive molecular and neuroanatomical characterization studies on the CRH system and the study of physiological and behavioural effects of acutely administered CRH, hardly any investigations have focussed on the neurobiological consequences of long-term central CRH dysfunction. The latter should be of prime concern given that chronic disturbances in this system are most likely to occur during, and to be involved in the development of, stress-related disorders. We have, therefore, developed an animal model of central CRH hyperactivity, i.e., the long-term icv CRH-infused rat. Rats receive CRH icv for seven days via a miniosmotic pump that is positioned underneath the skin and connected to an icv cannula via polyethylene tubing (Labeur et al. 1995; Linthorst et al. 1997; Reul et al. 1998). This animal model enables the study of the consequences of CRH hyperactivity for physiological, immune, neurochemical and behavioural processes under baseline conditions and after stressful challenges. Because CRH plays a major role in neuroimmunoendocrine interactions (Irwin et al. 1987; Rivier and Rivest 1993) and changes in the immune system have been described in major depression (Maes 1995; Sluzewska et al. 1996), we embarked on determing the effects of peripheral inflammation in the long-term icv CRH-treated rat.

The Long-Term icv CRH-Infused Rat: Changes Under Basal Conditions an Following a Peripheral Inflammatory Challenge

Immune System Changes

In a first study, the effects of long-term icv CRH infusion on the immune system were investigated (Labeur et al. 1995). It was observed that the spleen and thymus of CRH-infused rats were markedly involuted. To investigate changes in cellular immunity, in vitro T-cell proliferation was tested in primary spleen lymphocytes collected at day seven of the CRH treatment. As expected, lymphocytes of long-term CRH-treated animals showed suppressed proliferative responses to the mitogen concanavalin A. Surprisingly, however, levels of the main T-cell growth factor, IL-2, were elevated in the supernatant of spleen lymphocyte cultures of CRH-infused rats, despite the high levels of circulating glucocorticoid hormones in this animal model (see below). The elevated IL-2 levels may be the result of increased synthesis and/or accumulation of the growth factor due to reduced internalization because of the decreased expression of high-affinity IL-2 receptors on the T-cells. The latter may be the real reason for the decreased T-cell growth, because a reduced gene expression of the IL-2Rα chain was indeed observed in splenic T-cells of CRH-treated rats (Labeur et al. 1995; Reul et al. 1998). The effects of the CRH treatment on proliferative responses and the IL-2 system were abolished when the animals were adrenalectomized at the start of peptide infusion (Labeur et al. 1995) suggesting an intermediary role for the adrenal gland.

In subsequent experiments, effects of the CRH treatment on proinflammatory cytokines were investigated. Glucocorticoid hormones are well-known inhibitors of inflammatory response and this is generally regarded to be the result of their suppressive action on the synthesis of a variety of proinflammatory cytokines such as IL-1, IL-6, TNF-α, and other mediators. Surprisingly, however, splenocytes of the hypercortisolemic CRH-treated rats showed constitutively elevated levels of IL-1β mRNA expression. Moreover, when spleen lymphocytes or spleen macrophages were cultured and stimulated in vitro with LPS, cells of CRH-treated rats produced significantly higher levels of IL-1β transcripts than those of control animals (Labeur et al. 1995). Thus, it seems that the IL-1 system is hyper-responsive in long-term icv. CRH-infused rats, a finding that is underscored by in vivo experiments on the effects of LPS on proinflammatory cytokines. When rats were injected ip with LPS on day seven of the CRH treatment, their circulating levels of bioactive IL-1 and IL-6 were much higher than those in LPS-injected control animals (Linthorst et al. 1997). There were no differences in the TNF responses. These observations show that, under inflammatory conditions, the chronically CRH-treated rats show hypersecretion of the proinflammatory cytokines IL-1 and IL-6. On the one hand, this may be regarded as a compensatory mechanism to counterbalance the elevated levels of glucocorticoids, which are well known to suppress the expression and action of many cytokines (Munck and Guyre 1991). On the other hand, given the involvement of IL-1β in the physiological and pathological changes associated with inflammation and sepsis, this situation may be potentially dangerous for the organism (Dinarello 1991).

These data elegantly show that in vivo models of glucocorticoid hypersecretion may lead to effects that are totally in contrast with the ones expected on the basis of data generated solely by in vitro experiments. With regard to the long-term icv CRH infusion model, important characteristics may be the enhanced output of both the HPA axis and, for part of the time (see following), the sympathetic nervous system, a combination that could be of ultimate significance for the changes in immune system function in this model. These physiological features may make this model more compatible with certain chronic stress states than those that only implement in vivo treatment with high doses of glucocorticoid hormones.

HPA Axis Changes

Continuous infusion of CRH into the brain ventricular system resulted in sustained activation of the HPA axis, as revealed by the elevated plasma levels of ACTH and corticosterone measured on day seven of infusion (Labeur et al. 1995). Furthermore, in vivo microdialysis experiments conducted on the seventh day of infusion showed that the free (biologically available) levels of corticosterone were also increased in CRH-infused rats, which resulted in an abrogation of the normal diurnal rhythm in this glucocorticoid hormone. In conditions such as chronic stress and major depression, a similar disappearance of the diurnal glucocorticoid rhythm has been observed (Owens and Nemeroff 1991; Holsboer and Barden 1996). The state of a continuously hyperactive HPA axis is also underlined by the adrenal enlargement, thymus involution and increased expression of proopiomelanocortin (POMC) mRNA in the anterior pituitary of CRH-treated rats (Labeur et al. 1995; Linthorst et al. 1997).

Intraperitoneal injection of LPS elicited in both icv CRH-infused and control rats an activation of the HPA axis, as indicated by elevated free corticosterone levels in the dialysates (Fig. 3C). Although the attained peak levels in the two LPS-injected groups were similar, there was a significant delay before the free corticosterone started to rise in the long-term CRH-infused animals (Linthorst et al. 1997). Thus, despite the largely elevated levels of circulating glucocorticoids and the (very likely (see below)) CRH receptor desensitization, it appears that the HPA axis in long-term CRH-infused rats is still able to respond to an inflammatory challenge. Nevertheless, the pathophysiological consequence of such delayed corticosterone responses may be that glucocorticoid-facilitated and glucocorticoid-regulated processes are executed (too) late (Wiegers et al. 1995; Wiegers and Reul 1998).

Changes in Body Temperature and Locomotion

To monitor the effects of continuous icv infusion of CRH on body temperature and locomotor activity, we applied a biotelemetry method (Linthorst et al. 1997). Continuous icv treatment with CRH increased body temperature during the light as well as the dark period of the diurnal cycle resulting in a flattened circadian rhythm, especially during the first three to four days of

→

Fig. 3. Effects of ip administration of LPS (100 μg/kg) on hippocampal extracellular levels of 5-HT (**A**), behavioural activity (**B**), free corticosterone (**C**) and body temperature (**D**) in long-term icv vehicle-infused (△) and icv CRH-infused rats (●). In **A**, * and + indicate the first time point of statistical significance (as compared to ip saline-injected animals, not shown here) for icv vehicle- and CRH-infused rats, respectively (Student's t test). In **B** and **C**, * indicates the time period during which data from long-term icv vehicle- and CRH-treated animals differ significantly (MANOVA with repeated measures). In **D**, * indicates a significant difference between vehicle- and CRH-infused animals (Student's t test). The time point of injection is indicated by the arrow. The shaded area in **D** represents the dark period of the diurnal cycle. Data taken from Linthorst et al (1997) with permission from the Journal of Neuroscience

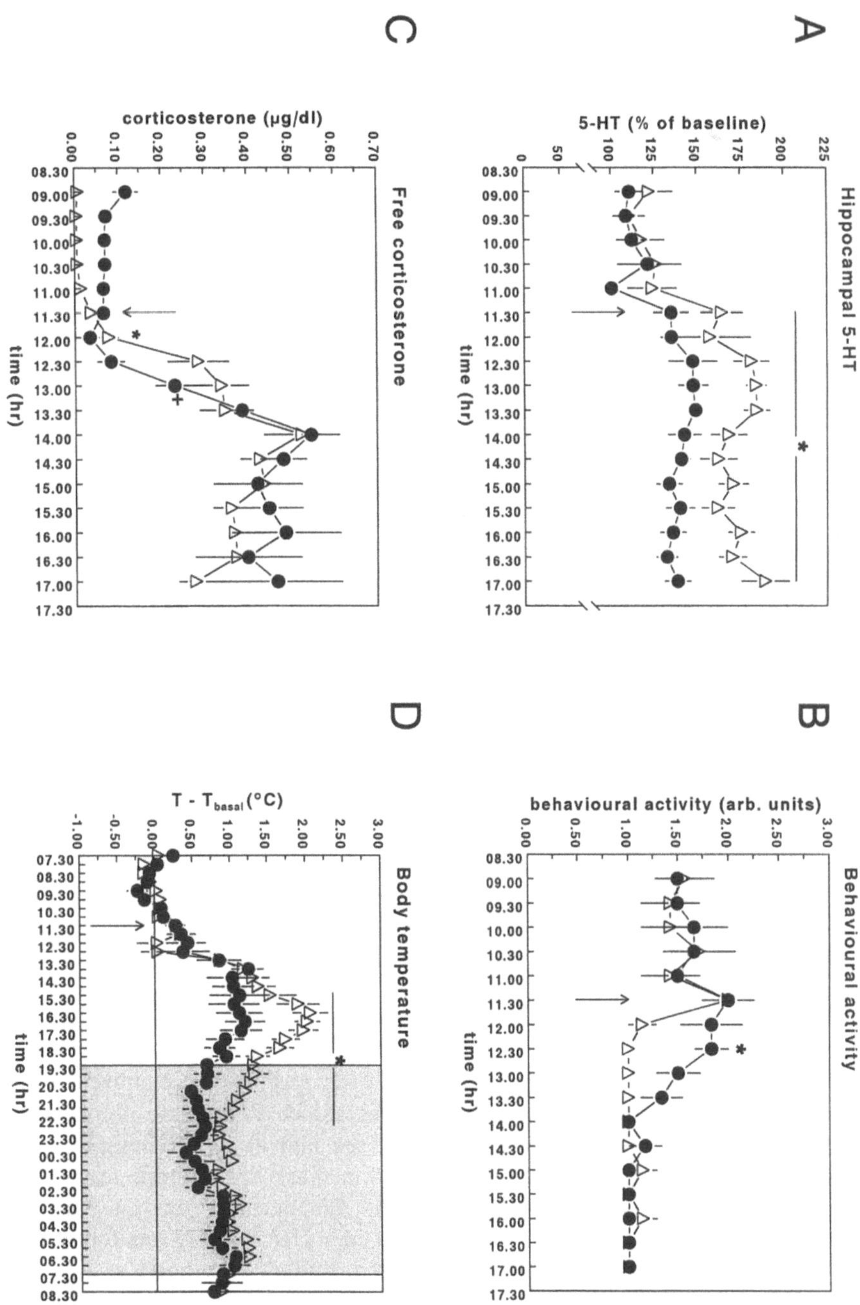
A
Hippocampal 5-HT
5-HT (% of baseline)
time (hr)
B
Behavioural activity
behavioural activity (arb. units)
time (hr)
C
Free corticosterone
corticosterone (µg/dl)
time (hr)
D
Body temperature
T - Tbasal (°C)
time (hr)

treatment. Moreover, the icv treatment enhanced locomotor activity during the light and the dark period of the first two to three days (Linthorst et al. 1997). The disappearance of the CRH-evoked increments in body temperature and locomotion over the treatment period indicates that central CRH effector mechanisms become desensitized by the continuous CRH load. Such a desensitization may involve changes in CRH receptor levels, as various conditions that are known to raise brain CRH levels, including chronic stress and adrenalectomy, decrease CRH receptor density and CRH mRNA concentrations in the anterior pituitary, the hypothalamus and/or the frontal cortex (Anderson et al. 1993; De Souza et al. 1985; Hauger et al. 1988; Luo et al. 1995; Makino et al. 1995; Tizabi and Aguilera 1992).

Induction of a peripheral inflammatory response by ip injection of LPS elicited a pronounced fever response in the control rats, whereas this response was highly attenuated in the long-term icv CRH-treated animals (Fig. 3D). However, long-term *peripherally* CRH-infused rats showed normal LPS-induced fever responses. It should be noted, that peripherally and icv CRH-infused animals showed similar elevations in free corticosterone levels under baseline conditions. Thus, the unaltered fever response in peripherally CRH-treated rats suggests that elevated levels of corticosterone are not responsible for the attenuated fever observed in long-term icv CRH-infused animals (Linthorst et al. 1997). These results underscore that a central action of CRH was responsible for the attenuated LPS-induced fever response in the icv CRH-treated animals.

Hippocampal Serotonergic Neurotransmission and Behavioural Activity

After seven days of icv CRH infusion, no changes in baseline extracellular levels of 5-HT and 5 HIAA in the hippocampus were observed as revealed by in vivo microdialysis (Linthorst et al. 1997). This was a surprising observation, because *acute* icv administration of CRH was shown to exert a marked increase in both 5-HT and 5-HIAA levels in the hippocampus (Reul and Linthorst 1997). Thus, it seems that long-term icv CRH treatment results in pertinent regulatory changes in 5-HT release patterns, an effect possibly related to the CRH-evoked desensitization of CRH receptors (see above).

Previously, ip administration of LPS was shown to increase hippocampal extracellular levels of 5-HT and 5-HIAA (see above). When long-term icv CRH-infused rats were inoculated with LPS, a markedly reduced response in hippocampal 5-HT was observed (Fig. 3A). Furthermore, in line with this blunted 5-HT response, a delayed elevation in 5-HIAA levels was found after ip administration of LPS in icv CRH-treated animals (Linthorst et al. 1997). These findings clearly indicate that a continuously elevated drive of CRH in the brain results in a blunted responsiveness of the hippocampal 5-HT system to an acute stressful challenge. Whether CRH receptor desensitization or putative alterations in brain cytokines (IL-1?) or both are underlying the reduced responsiveness in 5-HT needs to be further elucidated.

With regard to the behaviourial implications of ip-injected LPS during a condition of chronically elevated brain CRH levels, the animals did develop the typical signs of sickness behaviour (e.g., hypolocomotion) but with a delay of about 1.5-2 hr (Fig. 3B; Linthorst et al. 1997). As sickness behaviour is an important adaptive response supporting the host defence response, delayed initiation of this behaviour presents a potentially detrimental situation for the organism.

Consequences of Long-Term Hyperactivity of CRH in the Brain

In this review, we have demonstrated that a continuously elevated CRH drive in the brain results in marked changes in a variety of physiological systems. Notably, these changes resemble several characteristics seen in major depression. Hence, the CRH-infused rats show elevated levels of corticosterone, being devoid of an apparent circadian rhythm. Moreover, for the first time it was shown that aberrations in serotonergic neurotransmission can evolve from chronic CRH hyperactivity, a condition that may also develop in stress-related disorders such as major depression. The overall immunosuppression in combination with the hyperresponsiveness of IL-1 and IL-6 levels during peripheral inflammation indicates that a condition of chronically increased levels of CRH in the brain exerts distinct influences on immune system functioning. Yet the mechanisms underlying the chronic CRH-evoked alterations at the central nervous system and immune system levels still need to be elucidated, but appear to include CRH receptor desensitization, chronic HPA hormone hypersecretion and elevated sympathetic outflow.

Concluding Remarks

The immune system, the endocrine system and the CNS build a complex network of multiple interactions at various levels of biological organization. We have reviewed evidence that, during peripheral inflammation activation of hippocampal serotonergic neurotransmission occurs. Furthermore, the proinflammatory cytokine IL-1 appears to play an important intermediary role in inflammation-induced serotonin responses in the brain. Although under normal conditions the neuroimmunoendocrine network is thought to be able to respond adequately to disturbances of homeostasis, we have demonstrated that chronic elevation of brain CRH drive, a condition also occurring during chronic stress and stress-related disorders, results in aberrant immune, neuroendocrine, neurochemical, autonomic and behavioural responses to inflammation. Thus, chronically elevated levels of brain CRH might compromise biological defence mechanisms to cope with stressful challenges. Ultimately, such defective responses may have deleterious consequences for homeostasis, promote disease and limit the chance of survival of the organism.

Acknowledgements. The authors thank Ms. B. Burkart-Lauer for excellent artwork and Ms. E. Fesl for reading the manuscript. The help of Dr. S. Pauli with the $5HT_{1A}$ receptor autoradiography is also acknowledged. The work of the authors described in this chapter was subsidized by the Volkswagen Foundation (grant nos. 1/68 430 and I/70 543) and by the BIOMED Concerted Action Program "Cytokines in the brain".

References

Anderson SM, Kant GJ, De Souza EB (1993) Effects of chronic stress on anterior pituitary and brain corticotropin-releasing factor receptors. Pharmacol Biochem Behav 44:755–761

Auerbach SB, Minzenberg MJ, Wilkinson LO (1989) Extracellular Serotonin and 5-hydroxyindoleacetic acid in hypothalamus of the unanesthetized rat measured by in vivo dialysis coupled to high-performance liquid chromatography with electrochemical detection: dialysate serotonin reflects neuronal release. Brain Res 499:281–290

Brady LS, Lynn AB, Nerkenham M, Gottesfeld Z (1994) Systemic interleukin-1 induces early and late patterns of c-fos mRNA expression in brain. J Neurosci 14:4951

Connor TJ, Song C, Leonard BE, Merali Z, Anisman H (1998) An assessement of the effects of central interleukin-1-beta, -2, -6, and tumor necrosis factor-alpha administration on some behavioural, neurochemical, endocrine and immune parameters in the rat. Neuroscience 84:923–933

Crespi F, Garratt JC, Sleight AJ, Marsden CA (1990) In vivo evidence that 5-hydroxytryptamine (5-HT) neuronal firing and release are not necessarily correlated with 5-HT metabolism. Neuroscience 35:139–144

Dantzer R, Bluthé RM, Kent S, Kelley KW (1991) Behavioural effects of cytokines. In: Rothwell N, Dantzer R (eds) Interleukin-1 in the brain. Pergamon Press, Oxford, pp 135–150

De Souza EB, Insel TR, Perrin MH, Rivier J, Vale WW, Kuhar MJ (1985) Differential regulation of corticotropin-releasing factor receptors in anterior and intermediate lobes of pituitary and in brain following adrenalectomy in rats. Neurosci Lett 56:121–128

De Souza EB, Whitehouse PJ, Kuhar MJ, Price DL, Vale WW (1986) Reciprocal changes in corticotropin-releasing factor (CRF)-like immunoreactivity and CRF receptors in cerebral cortex of Alzheimer's disease. Nature 319:593–595

Dijkstra I, Tilders FJH, Aguilera G, Kiss A, Rabadan-Diehl C, Barden N, Karanth S, Holsboer F, Reul JMHM (1998) Reduced activity of hypothalamic corticotropin-releasing hormone neurons in transgenic mice with impaired glucocorticoid receptor function. J Neurosci 18:3909–3918

Dinarello CA (1991) Interleukin-1 and interleukin-1 antagonism. Blood 77:1627–1652

Dunn AJ (1992) Endotoxin-induced activation of cerebral catecholamine and serotonin metabolism: comparison with interleukin-1. J Pharmacol Exp Ther 261:964–969

Dunn AJ, Welch J (1991) Stress- and endotoxin-induced increases in brain tryptophan and serotonin metabolism depend on sympathetic nervous system activity. J Neurochem 57:1615–1622

Elmquist JK, Ackermann MR, Register KB, Rimler RB, Ross LR, Jacobson CD (1993) Induction of fos-like immunoreactivity in the rat brain following pasteurella-multocida endotoxin administration. Endocrinology 133:3054–3057

Ericsson A, Kovacs KJ, Sawchenko PE (1994) A functional anatomical analysis of central pathways subserving the effects of interleukin-1 on stress-related neuroendocrine neurons. J Neurosci 14:897–913

Gottschall PE, Komaki G, Arimura A (1992) Interleukin- 1β activation of the central nervous system. In: Rothwell N, Dantzer R (eds) Interleukin-1 in the brain. Pergamon Press, Oxford, pp 27–49

Hart BL (1988) Biological basis of the behavior of sick animals. Neurosci Biobehav Rev 12:123–137

Hauger RL, Millan MA, Lorang M, Harwood JP, Aguilera G (1988) Corticotropin-releasing factor receptors and pituitary adrenal responses during immobilization stress. Endocrinology 123: 396–405

Holsboer F, Barden N (1996) Antidepressants and hypothalamic pituitary adrenocortical regulation. Endocrinol Rev 7:187–205
Irwin MR, Vale W, Britton KT (1987) Central corticotropin-releasing factor suppresses natural killer cytotoxicity. Brain Behav Immun 1:81–87
Irwin M, Lacher U, Caldwell C (1992) Depression and reduced natural killer cytotoxicity: a longitudinal study of depressed patients and control subjects. Psychol Med 22:1045–1050
Jacobs BL, Azmitia EC (1992) Structure and function of the brain serotonin system. Physiol Rev 72:165–229
Janssen RAJ, Mulder NH, The TH, Deleij L (1994) The immunobiological effects of interleukin-2 in vivo. Cancer Immunol Immunother 39:207–216
Kabiersch A, Del Rey A, Honegger CG, Besedovsky HO (1988) Interleukin-1 induces changes in norepinephrine metabolism in the rat brain. Brain Behav Immun 2:267–274
Kalén P, Strecker RE, Rosengren E, Björklund A (1988) Endogenous release of neuronal serotonin and 5-hydroxyindoleacetic acid in the caudate-putamen of the rat as revealed by intracerebral dialysis coupled to high-performance liquid chromatography with fluorimetric detection. J Neurochem 51:1422–1435
Kalén P, Rosegren E, Lindvall O, Björklund A (1989) Hippocampal noradrenaline and serotonin release over 24 hours as measured by the dialysis technique in freely moving rats: correlation to behavioural activity state, effect of handling and tail-pinch. Eur J Neurosci 1:181–188
Karanth S, Lyson K, McCann SM (1993) Role of nitric oxide in interleukin 2-induced corticotropin-releasing factor release from incubated hypothalami. Proc Natl Acad Sci USA 90:3383–3387
Katsuura G, Gottschall PE, Dahl RR, Arimura A (1988) Adrenocorticotropin release induced by intracerebroventricular injection of recombinant human interleukin-1 in rats: possible involvement of prostaglandin. Endocrinology 122:1773–1779
Katsuura G, Gottschall PE, Dahl RR, Arimura A (1989) Interleukin-1 beta increases prostaglandin E2 in rat astrocyte cultures: modulatory effect of neuropeptides. Endocrinology 124:3125–3127
Kluger MJ (1991) Fever: role of pyrogens and cryogens. Physiol Rev 71:93–127
Komaki G, Arimura A, Koves K (1992) Effect of intravenous injection of IL-1 beta on PGE2 levels in several brain areas as determined by microdialysis. Am J Physiol 262:E246–E251
Labeur MS, Arzt E, Wiegers GJ, Holsboer F, Reul JMHM (1995) Long-term intracerebroventricular corticotropin-releasing hormone administration induces distinct changes in rat splenocyte activation and cytokine expression. Endocrinology 136:2678–2688
Lavicky J, Dunn AJ (1995) Endotoxin administration stimulates cerebral catecholamine release in freely moving rats as assessed by microdialysis. J Neurosci Res 40:407–413
Linthorst ACE, Flachskamm C, Holsboer F, Reul JMHM (1994) Local administration of recombinant human interleukin-1 beta in the rat hippocampus increases serotonergic neurotransmission, hypothalamic-pituitary-adrenocortical axis activity, and body temperature. Endocrinology 135:520–532
Linthorst ACE, Flachskamm C, Holsboer F, Reul JMHM (1995a) Intraperitoneal administration of bacterial endotoxin enhances noradrenergic neurotransmission in the rat preoptic area: relationship with body temperature and hypothalamic-pituitary-adrenocortical axis activity. Eur J Neurosci 7:2418–2430
Linthorst ACE, Flachskamm C, Müller-Preuss P, Holsboer F, Reul JMHM (1995b) Effect of bacterial endotoxin and interleukin-1 beta on hippocampal serotonergic neurotransmission, behavioral activity, and free corticosterone levels: an in vivo microdialysis study. J Neurosci 15:2920–2934
Linthorst ACE, Flachskamm C, Holsboer F, Reul JMHM (1996) Activation of serotonergic and noradrenergic neurotransmission in the rat hippocampus after peripheral administration of bacterial endotoxin: involvement of the cyclo-oxygenase pathway. Neurosci 72:989–997
Linthorst ACE, Flachskamm C, Hopkins SJ, Hoadley ME, Labeur MS, Holsboer F, Reul JMHM (1997) Long-term intracerebroventricular infusion of corticotropin-releasing hormone alters neuroendocrine, neurochemical, autonomic, behavioral, and cytokine responses to a systemic inflammatory challenge. J Neurosci 17:4448–4460
Luo X, Kiss A, Rabadandiehl C, Aguilera G (1995) Regulation of hypothalamic and pituitary corticotropin-releasing hormone receptor messenger ribonucleic acid by adrenalectomy and glucocorticoids. Endocrinology 136:3877–3883

Maes M (1995) Evidence for an immune response in major depression: a review and hypothesis. Prog Neuro-Psychopharmacol Biol Psychiat 19:11–38
Makino S, Schulkin J, Smith MA, Pacak K, Palkovits M, Gold PW (1995) Regulation of corticotropin-releasing hormone receptor messenger ribonucleic acid in the rat brain and pituitary by glucocorticoids and stress. Endocrinology 136:4517–4525
Merali Z, Lacosta S, Anisman H (1997) Effects of interleukin-1-beta and mild stress on alterations of norepinephrine, dopamine and serotonin neurotransmission – a regional microdialysis study. Brain Res 761:225–235
Molina-Holgado F, Guaza C (1996) Endotoxin administration induced differential neurochemical activation of the rat brain stem nuclei. Brain Res Bull 40:151–156
Munck A, Guyre PM (1991) Glucocorticoids and immune function. In: Ader R, Felten DS, Cohen N (eds) Psychoneuroimmunology. Academic Press, San Diego, pp 447–474
Munck A, Guyre PM, Holbrook NJ (1984) Physiological functions of glucocorticoids in stress and their relation to pharmacological actions. Endocrinol Rev 5:25–44
Nagayama H, Lu JQ (1997) Circadian rhythm in the response to intracerebroventricular administration of 8-OH-DPAT. Brain Res 756:92–95
Nemeroff CB, Widerlov E, Bissette G, Walleus H, Karlsson I, Eklund K, Kilts CD, Loosen PT, Vale W (1984) Elevated concentrations of CSF corticotropin-releasing factor-like immunoreactivity in depressed patients. Science 226:1342–1344
Owens MJ, Nemeroff CB (1991) Physiology and pharmacology of corticotropin-releasing factor. Pharmacol Rev 43:425–473
Owens MJ, Nemeroff CB (1993) The role of corticotropin-releasing factor in the pathophysiology of affective and anxiety disorders: laboratory and clinical studies. In: Chadwick DJ, Marsh J, Ackrill K (eds.) Corticotropin releasing factor. Ciba Foundation Symposium 172. John Wiley, Chichester, pp 296–308
Pauli S, Linthorst ACE, Reul JMHM (1998) Tumour necrosis factor-alpha and interleukin-2 differentially affect hippocampal serotonergic neurotransmission, behavioural activity, body temperature and hypothalamic-pituitary-adrenocortical axis activity in the rat. Eur J Neurosci 10:868–878
Pomara N, Singh RR, Deptula D, Le Witt PA, Bissette G, Stanley M, Nemeroff CB (1989) CSF corticotropin-releasing factor (CRF) in Alzheimer's disease: its relationship to severity of dementia and monoamine metabolites. Biol Psychiat 26:500–504
Raadsheer FC, Hoogendijk WJG, Stam FC, Tilders FJH, Swaab DF (1994) Increased numbers of corticotropin-releasing hormone expressing neurons in the hypothalamic paraventricular nucleus of depressed patients. Neuroendocrinology 60:436–444
Raadsheer FC, Van Heerikhuize JJ, Lucassen PJ, Hoogendijk WJG, Tilders FJH, Swaab DF (1995) Corticotropin-releasing hormone mRNA levels in the paraventricular nucleus of patients with Alzheimer's disease and depression. Am J Psychiat 152:1372–1376
Raber J, Bloom FE (1994) IL-2 induces vasopressin release from the hypothalamus and the amygdala: Role of nitric oxide-mediated signaling. J Neurosci 14:6187–6195
Reul JMHM, Linthorst ACE (1997) Central administration of corticotropin-releasing hormone profoundly stimulates rat hippocampal serotonergic neurotransmission and behavioral activity: no effect of adrenalectomy Society Neurosci 56.7.
Reul JMHM, Labeur MS, Wiegers GJ, Linthorst ACE (1998) Altered neuroimmunoendocrine communication during a condition of chronically increased brain corticotropin-releasing hormone drive. Ann NY Acad Sci 840:444–455
Rivier C, Rivest S (1993) Mechanisms mediating the effects of cytokines on neuroendocrine functions in the rat. In: Chadwick DJ, Marsh K, Ackrill J (eds) Corticotropin-releasing factor. Ciba Foundation Symposium 172. John Wiley, Chichester, pp 204–225
Rivier C, Shen GH (1994) In the rat, endogenous nitric oxide modulates the response of the hypothalamic-pituitary-adrenal axis to interleukin-1 beta, vasopressin, and oxytocin. J Neurosci 14:1985–1993
Rothwell NJ (1991) Functions and mechanisms of interleukin 1 in the brain. Trends Pharmacol Sci 12:430–436
Rueter LE, Jacobs BL (1996) A microdialysis examination of serotonin release in the rat forebrain induced by behavioral environmental manipulations. Brain Res 739:57–69

Sagar SM, Price KJ, Kasting NW, Sharp FR (1995) Anatomic patterns of FOS immunostaining in rat brain following systemic endotoxin administration. Brain Res Bull 36:381–392

Sirko S, Bishai I, Coceani F (1989) Prostaglandin formation in the hypothalamus in vivo: effect of pyrogens. Am J Physiol 256:R616–624

Sluzewska A, Rybakowski J, Bosmans E, Sobieska M, Berghmans R, Maes M, Wiktorowicz K (1996) Indicators of immune activation in major depression. Psychiat Res 64:161–167

Smith T, Hewson AK, Quarrie L, Leonard JP, Cuzner ML (1994) Hypothalamic PGE(2) and cAMP production and adrenocortical activation following intraperitoneal endotoxin injection – in vivo microdialysis studies in Lewis and Fischer rats. Neuroendocrinology 59:396–405

Tizabi Y, Aguilera G (1992) Desensitization of the hypothalamic-pituitary-adrenal axis following prolonged administration of corticotropin-releasing hormone or vasopressin. Neuroendocrinology 56:611–618

Vale W, Vaughan J, Perrin M (1997) Corticotropin-releasing factor (CRF) family of ligands and their receptors. Endocrinologist 7:S3–S9

Wan W, Janz L, Vriend CY, Sorensen CM, Greenberg AH, Nance DM (1993) Differential induction of c-Fos immunoreactivity in hypothalamus and brain stem nuclei following central and peripheral administration of endotoxin. Brain Res Bull 32:581–587

Wiegers GJ, Reul JMHM (1998) Induction of cytokine receptors by glucocorticoids: functional and pathological significance. Trends Pharmacol Sci 19:317–321

Wiegers GJ, Labeur MS, Stec IEM, Klinkert WEF, Holsboer F, Reul JMHM (1995) Glucocorticoids accelerate anti-T cell receptor-induced T cell growth. J Immunol 155:1893–1902

Zalcman S, Green-Johnson JM, Murray L, Nance DM, Dyck D, Anisman H, Greenberg AH (1994) Cytokine-specific central monoamine alterations induced by interleukin-1, -2 and -6. Brain Res 643:40–49

Neuronal Control of the Immune Response in the Central Nervous System: From Pathogenesis to Therapy

H. Wekerle, A. Flügel, and H. Neumann

Introduction

Immune surveillance of the central nervous System (CNS) is conditional. It is minimal in the state of health, but can be strongly upregulated whenever required. A number of structural features are responsible for this peculiar type of immune reactivity. First, the CNS tissues are secluded from the rest of the body by a tight endothelial blood-brain barrier (BBB). This barrier is impermeable for most blood-borne macromolecules and cells. Then, the CNS tissues are not drained by a fully organized lymphatic vasculature, which in most other tissues ensures the transport of antigenic material to the neighboring lymphoid organs. Finally, the CNS lacks professional antigen-presenting cells, the sentinels, which process antigenic material to become recognizable by recirculating specific T lymphocytes. In fact, the healthy CNS shows a general deficit of major histocompatibility complex (MHC) antigens.

But this apparent lack of immune reactivity ("immune privilege") applies only to the *healthy* CNS. In the course of pathological processes, a gene program is turned on that changes the non-reactive CNS tissue to a pro-inflammatory milieu, which aptly supports vigorous immune reactions. This implies the induction of MHC expression on many, if not all, brain cells, as well as the production of soluble mediators, such as chemokines and cytokines.

It has become evident over the past few years that a remarkably broad spectrum of pathological changes is able to elicit immune reactivity in the CNS. These include, beyond immune-mediated inflammation, microbial infections, neoplastic transformation, and in particular neurodegenerative changes. Parkinson's and Alzheimer's diseases, Huntington's chorea, amyotrophic lateral sclerosis and adrenoleukodystrophy are prominent examples of neurodegeneration coinciding with the induction of MHC antigens and cytokines, and, in some cases, florid invasion of inflammatory cells (Neumann and Wekerle 1998). This article first summarizes recent work aimed at elucidating the relationship between neuronal activity (and its lack) with immune reactivity in the CNS. As it turns out, functional activity of neurons strongly suppresses immune reactivity (as evidenced by MHC expression) both in surrounding glia cells and in the neurons themselves. Paralysis of neurons by neurotoxins unleashes expression of the pro-inflammatory genes, rendering the lesioned CNS areas attractive for immune cells.

Patterson/Kordon/Christen (Eds.)
Neuro-Immune Interactions in Neurologic and Psychiatric Disorders

A second part of the article describes therapeutic implications of these findings. A new therapeutic strategy to treat neurodegenerative diseases uses T lymphocytes as vehicles to transport therapeutically useful gene products through the BBB into CNS lesions. The approach uses recombinant retroviruses to introduce genes encoding neurotrophins or other protein mediators into brain-specific T lymphocytes. The engineered T cells are targeted to neurodegenerative CNS foci, where they produce and deposit their neurotrophic "payload," with the intention of rescuing and protecting the compromised local cell populations.

Neuronal Activity Controls Immune Responsiveness in Brain

Neuronal Transection Studies

A particularly rewarding model to study the impact of neuronal activity on the cellular milieu is based on unilateral clipping of the nervus facialis, a purely motor cranial nerve. As detailed by the studies from Kreutzberg's group, peripheral transection of the facial nerve has profound effects on the cells surrounding the perikarya located remotely in the brain stem. Synapses contacting lesioned neuronal bodies are pruned within hours by microglia in response to the loss of synaptic input (Kreutzberg 1996). This reorganization of neuronal connections shields lesioned neurons from synaptic activity and possibly allows the establishment of new functional synaptic connections. These events coincide with the de *novo* expression of MHC class I and class II molecules on the activated perineuronal glial cells (Kreutzberg 1996). As shown in other experimental systems, comparable changes are noted after axonal transection within the CNS, which abrogates normal physiological input to the innervated target tissue, dramatically enhancing local expression of MHC class I and class II molecules (Finsen et al. 1993).

Regulation of MHC expression by neuronal signals was also noted in the cases of skeletal muscles innervated by motor neurons. Blockade of neuronal input and muscular electrical activity by axotomy or tetrodotoxin treatment of the innervating nerve increased MHC expression in the muscle tissue. External electric stimulation of the transected distal nerve trunk reversed this upregulation of MHC molecules on the muscle fibers (Gundersen and Maehlen 1994).

Importantly, however, MHC expression is not the only immune-related function controlled by neuronal activity. In addition, local production and secretion of pro-inflammatory cytokines and chemokines are regulated tightly by neurons. Studies of the peripheral facial nerve axotomy system showed induction of local gene transcription of tumor necrosis factor-α (TNF-α), interleukin-1β (Il-1β) and interferon-γ (IFN-γ) in the facial nucleus (Raivich et al. 1998), whereas in another model, spinal cord lesions resulted in de novo expression of the chemokines MIP-1α and β (Bartholdi and Schwab 1997).

The lesion-dependent induction of MHC molecules, pro-inflammatory cytokines and chemotactic mediators should create a CNS milieu attractive for recirculating lymphocytes and supportive to their local reactivity. In fact, neuronal manipulations resulting in the appearance of MHC molecules in the CNS predictively further the invasion of T lymphocytes to areas containing the damaged neuronal perikarya (Raivich et al. 1998; Olsson et al. 1992). This is quite remarkable considering that the BBB is not disturbed in this area of immune activation. Lymphocytic invasion of neurodegenerative lesion areas is especially prominent following the transfer of in vitro activated T cells, for example, after i.v. injection of myelin basic protein-specific T cells into rats whose motoneurons (Maehlen et al. 1989) or optic nerves (Hickey 1991; Konno et al. 1990) had been cut before. Within hours, T cells leave the blood stream through the intact BBB to find their way to the CNS regions harboring the somata of the peripherally lesioned neurons.

Attraction of immune cells by neurodegenerative CNS areas seems to be a physiological phenomenon, by no means restricted to transferred, ex vivo activated T cells. In facial nerve axotomized mice, invasion of lesioned facial nucleus areas by polyclonal endogenous lymphocytes and macrophages is a normal feature (Raivich et al. 1998), and similar findings have been reported in spinal cord lesions (Schnell et al. 1997).

Taken together, these studies establish that immune responsiveness in the CNS is tightly controlled by neuronal signals. Following their functional inactivation, neurons as well as their neighboring cells become competent for interactions with the T lymphocytes, which patrol the nervous system parenchyma (Fig. 1).

In theory, there are at least two principal pathways by which neurons could regulate immune responsiveness. First, transection of axonal processes might activate gene transcription of immune mediators in neurons, which, in turn, would stimulate local CNS cells to express MHC molecules and to pre-

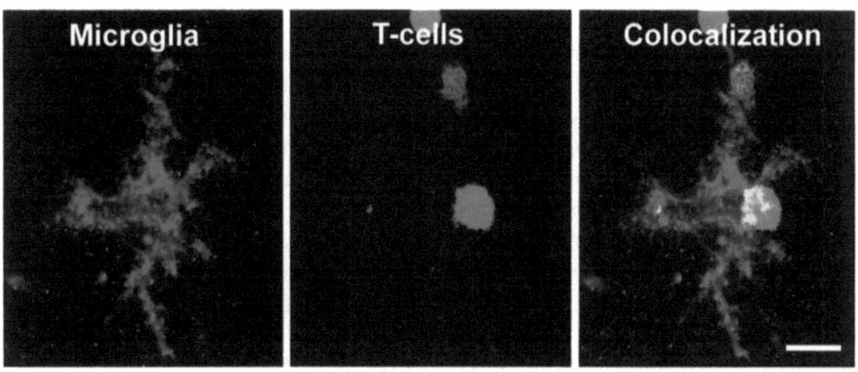

Fig. 1. Association of autoreactive T cells with microglia. MBP-specific CD4-positive T cells were transduced with GFP. T cells can be detected by confocal laser scanning microscopy in close proximity to immunolabeled microglia four days after T cell transfer in rats. Scale bar: 10 μm

sent antigen. This pathway is supported by a report describing IFN-γ immunoreactivity in facial motor neurons following axotomy (Olsson et al. 1989), although the molecular nature of this material was questioned by others (Kiefer et al. 1991). Moreover, this would not directly explain remote immune activation in the innervated target tissue, where changes are most likely a consequence of a lack of physiological neuronal input i.e., normal electrical and synaptic activity (Gundersen an Maehlen 1994).

Second, electrically active neurons could continuously suppress inducibility of pro-inflammatory gene expression. Loss of electrical or synaptic activity would then result in facilitated gene inducibility. Possible mechanisms participating in these regulatory events will be discussed below.

MHC Regulation by Neuronal Activity

The in vivo models of axonal transection studies indicated a potential suppressive effect of neurons on the immune status of nervous tissue, but they did not elucidate the nature of the neuronal signals that initiate and terminate the immune responsiveness. To approach these questions, we performed a series of in vitro studies using cultures of organotypic CNS explants and dissociated neural cells.

Regulation of MHC expression was examined in glial cells using a hippocampal slice culture. These brain slices preserve their organotypic structure and model the in vivo interplay of neural cell types in vitro. Within these tissue cultures, the neurons are embedded in glial cells and establish an interconnected network allowing exchange of spontaneous action potentials (Gähwiler 1988).

Analysis of this cultured brain tissue demonstrated that intact neurons suppress IFN-γ mediated induction of MHC class II molecules on surrounding glia cells. Blockade of neuronal activity with tetrodotoxin (TTX) restored MHC class II inducibility on astrocytes (Neumann et al. 1996).

Detailed confocal microscopy studies showed that, within the cultured brain slices, electrically active neurons down-regulated the MHC class II inducibility in perineuronal glial cells (Fig. 2), and that this suppression was lifted by blockade of neuronal activity with TTX (Neumann et al. 1998).

Control of MHC expression by neuronal activity is by no means restricted to the glia cells forming the perineuronal milieu, but holds true for the neurons themselves (Neumann et al. 1995). We analyzed gene transcription of MHC class I heavy chain, β2-microglobulin and the peptide transporter TAP1/TAP2 in cultured dissociated rat hippocampal neurons by single-cell RT-PCR (Neumann et al. 1997). Reminiscent of the explant cultures, the neurons growing in the dissociated cell cultures establish communicating networks with sustained spontaneous electric activity.

MHC class I and class I-associated genes were inducible by IFN-γ. Blockade of neuronal activity with tetrodotoxin significantly increased the percentage of neurons transcribing MHC class I, β2-microglobulin and TAP1/TAP2

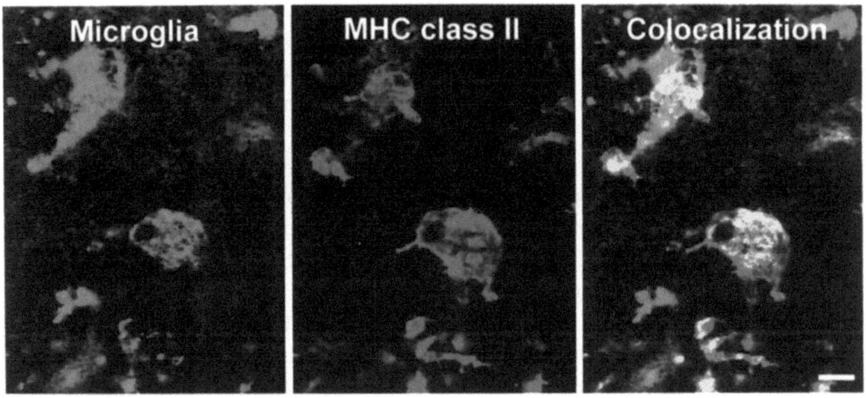

Fig. 2. Induction of MHC class II on microglia by IFN-γ. Cultured rat hippocampal slices were treated with IFN-γ (50 U/ml) for 72 hours. MHC class II molecules and microglia were immunolabeled and analyzed by confocal laser scanning microscopy. Scale bar: 10 µm

genes after IFN-γ treatment. Protein expression of MHC class I molecules on the neuronal cell membrane was analyzed by confocal laser scanning microscopy. MHC class I expression was inducible on dissociated cultured rat neurons by IFN-γ, but only in neurons that do not fire action potentials or after blockade of neuronal activity with TTX (Neumann et al. 1997a). Very similar results were reported by Lindå et al., who studied spinal motor neurons in situ, and also found a preferentially reduced gene transcription of β2-microglobulin (Lindå et al. 1998).

Recently, MHC class I gene transcripts were detected in CNS neurons of the cat and rat by in-situ hybridization. In this system, injection of TTX into the nervus opticus led to a decreased hybridization signal in the lateral geniculate nucleus (Corriveau et al. 1998). However, the in-situ hybridization approach used in this study does not allow distinction between classical and highly homologous non-classical MHC class I gene transcripts.

Mechanisms of Activity-Dependent MHC Regulation

Regulatory interactions between neurons and other cells in the local CNS environment could be mediated by 1) indirect electrical coupling due to change in ion concentrations, 2) cell adhesion molecules or 3) soluble mediators.

Electrical coupling. Membrane excitability is the hallmark of neurons. This type of electric activity involves dramatic changes in local ion concentrations, which have the potential to influence gene expression in neighboring cells via depolarization of their cell membrane. Such interactions could apply to astrocytes in particular, whose processes are in close contact with neurons and take up ions to buffer the local fluctuations in ion concentrations produced by electrically active neurons. This could trigger secondary signaling events in these cells, leading to altered expression patterns of molecules relevant for the local immune microenvironment. To date, however, direct evidence for this mechanism has been difficult to obtain and consequently is still lacking.

Membrane receptors. Adhesion molecules and membrane receptors could well be involved in the propagation of neuron-derived signals controlling the immune micro-environment. Its known that the expression of certain adhesion molecules is modulated by neuronal activity. The gene transcription of the adhesion molecule L1 as well as its translocation to the cell membrane are subject to such activity-dependent control, (Fields and Itoh, 1996). In addition, the polysialic acid form of the neural cell adhesion molecule (NCAM) is regulated by synaptic activity (Muller et al. 1996). Involvement of neuronal adhesion molecules in neuron-glial interactions has been demonstrated by many studies. But, again, there is presently no formal evidence for a role of neuronal adhesion molecules in regulating the immune microenvironment in the nervous system.

Soluble mediators. Neuronal electrical activity is directly linked to the release of a number of soluble mediators, including neurotransmitters and neurotrophins. Neurotransmitters, such as glutamate, can act on neighboring astrocytes, inducing Ca^{2+}-oscillations via specific ligand-gated receptors (Cornell-Bell et al. 1990; Dani et al. 1992). The stimuli mediated by neurotransmitters and neuropeptides could well affect immune functions of the target cells. Indeed, glutamate at high concentrations inhibits MHC class II-inducibility in astrocyte cultures (Lee et al. 1992). In our studies of cultured brain slices, down-regulation of MHC class-II in microglia cells by glutamate was noted, but it was the result of an indirect effect. Physiological concentrations of glutamate failed to affect MHC class II expression in isolated microglial cell cultures, (Neumann et al. 1998).

Other classical neurotransmitters and neuropeptides have been shown to reduce the MHC class II inducibility of cultured astrocytes. Norepinephrine and vasoactive intestinal polypeptide (VIP) acted directly on cultured astrocytes and prevented IFN-γ-mediated induction of MHC class II molecules (Frohman et al. 1988a, b).

We mentioned above that, in addition to neurotransmitters, neurotrophins are released from neurons dependent on electrical activity (Thoenen 1995; Lo 1995). Recently, we demonstrated that neurotrophins inhibit the MHC class II inducibility in microglia. In cultured brain slices, the neurotrophins nerve growth factor (NGF), brain-derived growth factor (BDNF) and neurotrophin-3 (NT-3) were independently capable of down-regulating inducibility of MHC class II on microglia, though to variable degrees. NGF and NT-3, but not BDNF, acted directly on isolated microglia culture and inhibited MHC class II expression mediated by IFN-γ (Neumann et al. 1998). The effect of BDNF on microglial MHC class II expression in tissue culture probably was an indirect one. It could be mediated principally by the release of NGF, whose release is stimulated by BDNF in neurons.

Of particular interest in this context are pro-inflammatory cytokines, which are known to modulate MHC expression and the local immunological milieu. In several studies, neurons have been shown to express pro-inflammatory cytokines like TNF-α. and IL-1 (Tehélingérian et al. 1993). We identified gene transcription and protein expression of IFN-γ in cultured sensory neurons by single cell RT-PCR and confocal laser scanning microscopy. IFN-γ acted back on its cellular source and activated the transcription factor STAT1 in sensory neurons (Neumann et al. 1997b). Whether neurons secrete pro-inflammatory cytokines in sufficient amounts to modulate the immune microenvironment remains to be shown.

Genetic Engineering of T-Cells: Therapy of Neurodegeneration?

As discussed above, the CNS is secluded from blood circulation by a tight endothelial BBB (Goldstein and Betz 1986). This BBB is beneficial for the CNS, holding back potentially dangerous material and ensuring the particular intercellular milieu required for proper brain function. On the other hand, the BBB forms an impermeable hurdle for most therapeutic macromolecules. For example, native neurotrophic factors show poor permeability through the intact BBB. Their delivery to CNS lesions requires manipulations of the BBB, coupling to BBB transporter systems (e.g., tansferrin), or direct, stereotactic deposition (Pardridge 1998). A therapeutic alternative is provided by activated T lymphocytes, which can cross the intact BBB and have a propensity to navigate to CNS lesions. We now show that brain-specific T lymphocytes genetically modified to produce relevant proteins can be used as vehicles to treat focal CNS lesions.

Gene Transfer into Antigen Specific $CD4^+$ T Lymphocytes

Gene delivery and stable expression of genes in native, non-transformed lymphocytes is generally not considered an easy task. Non-viral gene delivery techniques usually are very inefficient, possibly due to induction of apoptotic cell death (Ebert et al. 1997). Among the various viral delivery systems, replication-deficient retroviruses are currently the first choice for gene transfer into T lymphocytes, although in many cases they shut down expression of their transgene upon transfer in vivo (Verma and Somia 1997). While early studies of retroviral gene transfer into primary lymphocytes yielded transduction rates of around 1–10% (Kasid et al. 1990; Woffendin et al. 1994; Culver et al. 1991), co-cultivation of proliferating T lymphocytes with packaging lines increased the rate to 15–40%, (Mavilio et al. 1994).

We developed a highly efficient gene delivery technique that is based on the combination of three principles: 1) primary limiting dilution cultures.

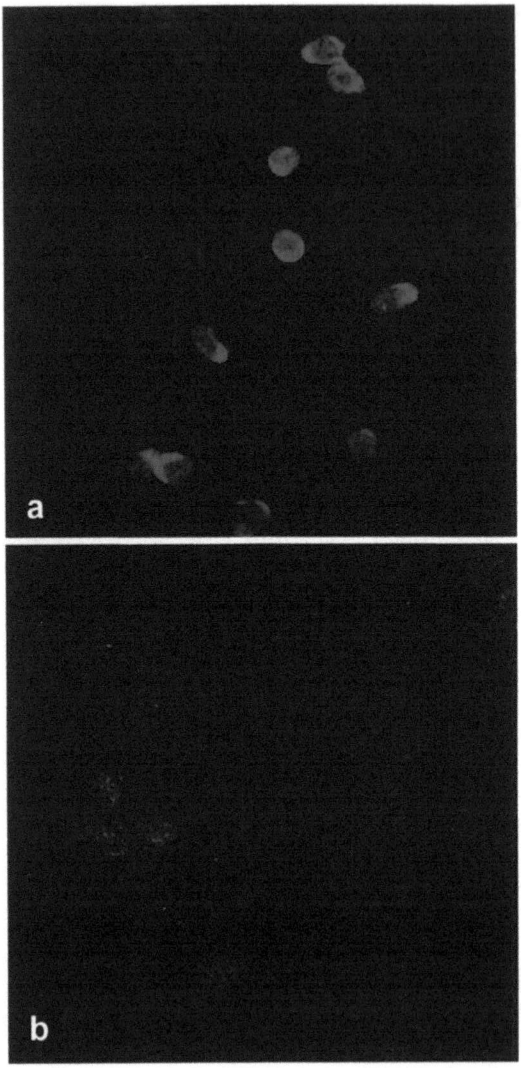

Fig. 3. a. P2-specific CD4-positive T lymphocytes transduced with NGF. Retroviral vector containing the NGF-gene was introduced into T lymphocytes with coculture/limited dilution transduction technique. The transgene expression was controlled following immunohistochemical staining with confocal laser scanning microscopy. b. Non-transduced P2-specific control T lymphocytes stained for NGF

This implies the establishment of a large number of parallel microcultures. Among these a few will eventually yield antigen-reactive, transduced T cell lines, which are composed of only one or a few clones. 2) co-culture of T cells with monolayers of packaging cells. The physical contact between T cells and packaging cells (which continuously release replication-deficient

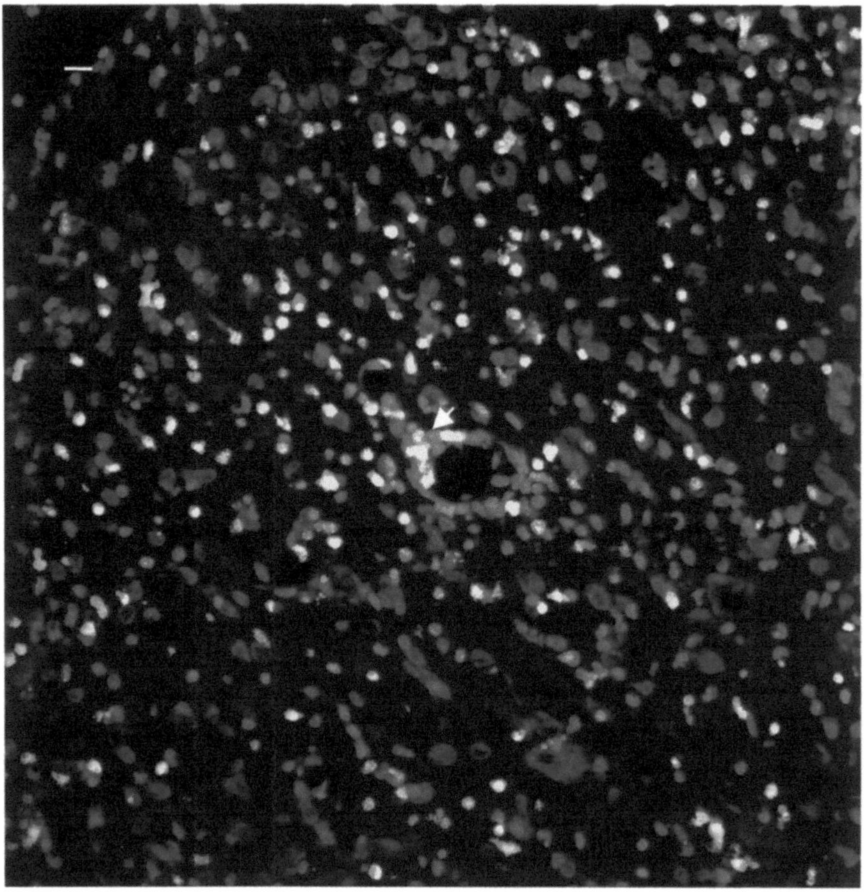

Fig. 4. GFP-transduced T cells in EAE lesions. MBP-specific T lymphocytes carrying the gene for GFP (T_{MBP}GFP cells) infiltrating the lumbar-sacral spinal cord four days following peripheral transfer are depicted. Green: GFP. Red: unspecific counterstain ToPro^{RM} -3 iodide. Yellow: double-staining with GFP and ToPro^{RM} -3 iodide. T_{MBP}GFP cells infiltrate perivascular areas (arrow) and the CNS parenchyma. Magnification bar: 10μm (From Flügel et al. 1999)

recombinant retroviruses) guarantees exposure of T cells to a maximal concentration of fresh retrovirus over a controlled time period. 3) simultaneous application of strong positive and negative selective pressures. Antigen-specific T cells are confronted with antigen-presenting cells, which display the relevant antigen. Recognition of the antigen triggers vigorous proliferation restricted exclusively to the antigen-specific T cell. At the same time, addition of a toxic antibiotic to the culture medium selects only those antigen-specific T cells that have incorporated the retrovirus. The recombinant retrovirus contains – besides the "payload" gene – a gene mediating resistance against the selecting antibiotic (Flügel et al. 1999). Autoantigen-specific T cells transduced with neurotrophic factors or the gene for green fluorescence protein (GFP) showed stable transgene expression in vitro and in vivo (Figs. 3, 4). Furthermore, they did not differ from their wild type colleagues in regard to their membrane phenotype and their (auto)immune function.

Gene Engineered T Lymphocytes as Therapeutic Vehicles

The first study exploiting tissue-specific delivery of therapeutic transgenes by T lymphocytes in the nervous system was described in 1995 by Kramer et al. (1995). Using the T cell transfer model of Experimental Autoimmune Neuritis (EAN) of the Lewis rat, they described the delivery of transgenic nerve growth factor (NGF) into peripheral nerves by genetically engineered, nerve-specific T lymphocytes. The locally deposited neurotrophin profoundly suppressed T cell-mediated local autoimmune disease, presumably via its anti-inflammatory effect described above (Neumann et al. 1998). NGF release in the target tissue drastically suppressed infiltration and activation of macrophages in the target tissue, resulting in almost complete preservation of periaxonal myelin sheaths. The histological effects were faithfully reflected by marked amelioration of clinical disease.

More recently, several studies applied a therapeutic transgene delivery via antigen-specific T lymphocytes to rodent models of experimental autoimmune encephalomyelitis (EAE). In one report, MBP-specific hybridoma cells were retrovirally transduced with an IL-4 construct. Transfer of the modified T lymphocytes was able to delay the onset and reduce the severity of actively induced EAE in (SJL/J×PL/J) F1 mice when administrated before the onset of disease (Shaw et al. 1997). A second study used lymph node cells of (SWR×SJL) F1 mice primed in vivo with proteolipid protein (PLP) peptide p139–151 to be transfected by the DMSO/polybrene technique with an IL-10 construct. The IL-10 engineered PLP-specific T cells protected the recipient mice against actively induced EAE when transferred before onset of clinical disease, and, in addition, were able to mitigate established disease (Mathisen et al. 1997). Further, mouse EAE was treated with T cells engineered to produce TGF-β as an anti-inflammatory mediator (Chen et al. 1999). The use of gene-engineered autoreactive T cells for the treatment of autoimmune disease was finally confirmed in additional disease models such as arthritis (Chernajovsky et al. 1997) and spontaneous, insulin-dependent diabetes mellitus of the NOD mouse (Moritani et al. 1996).

Conclusions

The finding that electric activity of neurons profoundly regulates immune reactivity in the CNS has multiple and important implications. Firstly, it indicates that immune reactivity in the CNS is under tight control of bipolar regulatory mechanisms. Neurons have a globally suppressive potential, with the intention to prevent and limit the development of inflammatory responses and their concomitant by stander damage. On the other hand, activated T cells crossing the BBB release potentially pro-inflammatory mediators. The balance between pro-inflammatory and suppressive elements determines localization, intensity and course of immune responses in the brain. In other words, an overwhelming invasion of activated T cells may create inflammation in intact parts of the CNS, whereas in lesioned areas, much weaker pro-inflammatory stimuli are required to create the same effect. In extreme case, in neurodegenerative tissues, "spontaneous" inflammatory infiltrates may be formed.

This finding should lead us to reevaluate the biological meaning of inflammatory cell infiltrates, which are so commonly seen in degenerative brain disorders. Are they good for the patient, or pernicious? Traditionally, any inflammatory response was considered a potential threat, dealing a final blow to the pre-damaged tissue components. However, the contrary may be the case. It has been realized very recently that normal inflammatory cells [T cells (Kerschensteiner et al. 1999); B cells (Torcia et al. 1996); and macrophages (Elkabes et al. 1996)] produce neurotrophic factors. On the other hand, pro-inflammatory cytokines, like TNF-α may have a role in regeneration of neural cells (Bruce et al. 1996). Infiltration of degenerative brain areas by inflammatory cells could thus reflect a beneficial process encouraging local cell regeneration.

This process can be enhanced by modern gene technology. Import of exogenous genes into brain-specific T cells, be it to enhance endogenous cytokine genes, or to introduce novel genes, appears to be a realistic approach to therapeutically enhance natural mechanisms of regeneration.

Acknowledgments. The original research discussed here is supported by grants from Deutsche Forschungsgemeinschaft (SFB-391 to H.N. and H.W., and SFB-455 to A.F. and H.W.), VW Foundation (to H.N. and H.W.), and European Community (CHRX-CT94-0670). We thank Dr. A. Blöchl for her substantial support in immunohistochemistry.

References

Bartholdi D, Schwab ME (1997) Expression of pro-inflammatory cytokine and chemokine mRNA upon experimental spinal cord injury in mouse: An in situ hybridization study. Eur J Neurosci 9:1422–1438

Bruce AJ, Boling W, Kindy MS, Peschon J, Kraemer PJ, Carpenter MK, Holtsberg FW, Mattson MP (1996) Altered neuronal and microglial responses to excitoxic and ischemic brain injury in mice lacking TNF receptors. Nature Med 2:788–794

Chen LZ, Hochwald GM, Huang C, Dakin G, Tao H, Cheng C, Simmons WJ, Dranoff G, Thorbecke GJ (1999) Gene therapy in allergic encephalomyelitis using myelin basic protein-specific T cells engineered to express latent transforming growth factor beta-1. Proc Natl Acad Sci USA 95: 12516–12521

Chernajovski Y, Adams G, Triantaphyllopoulos K, Ledda MF, Podhajcer OL (1997) Pathogenic lymphoid cells engineered to express TGF-β1 ameliorate disease in a collagen-induced arthritis. Gene Ther 4:553–559

Cornell-Bell AH, Finkbeiner S, Cooper MS, Smith SJ (1990) Glutamate induces calcium waves in cultured astrocytes: Long-range glial signaling. Science 247:470–473

Corriveau RA, Huh GS, Shatz CJ (1998) Regulating of class I MHC gene expression in the developing and mature CNS by neural activity. Neuron 21:505–520

Culver K, Cornetta K, Morgan R, Morecki S, Aebersold P, Kasid A, Lotze M, Rosenberg SA, Anderson WF, Blaese RM (1991) Lymphocytes as cellular vehicles for gene therapy in mouse an man. Proc Natl Acad Sci USA 88:3155–3159

Dani JW, Chernjavski A, Smith SJ (1992) Neuronal activity triggers calcium waves in hippocampal astrocyte networks. Neuron 8:429–440.

Ebert O, Finke S, Salahi A, Herrmann M, Trojaneck B, Lefterova P, Wagner E, Kircheis R, Huhn D, Schriever F, Schmidt-Wolf IG (1997) Lymphocyte apoptosis: Induction by gene transfer techniques. Gene Ther 4:296–302.

Elkabes S, DiCicco-Bloom EM, Black IB (1996) Brain microglia/macrophages express neurotrophins that selectively regulate microglial proliferation and fuction J Neurosci 16:2508–2521.

Fields RD, Itoh K (1996) Neural cell adhesion molecules in activity-dependent development an synaptic plasticity. Trends Neurosci 19:473–480.

Finsen BR, Tönder N, Xavier GF, Sörensen JC, Zimmer J (1993) Induction of microglial immunomolecules by anterogradely degenerating mossy fibers in the rat hippocampal formation. J Chem Neuronat 6:276–275.

Flügel A, Willem M, Berkowicz T, Wekerle H (1999) Gene transfer into $CD4^+$ T lymphoctes: Green fluorescent protein engineered, encephalitogenic T cells used to illuminate immune responses in the brain. Nature Med 5:843–847

Frohman EM, Vayuvegula B, Gupta S, Van den Noort S (1988a) Norepinephrine inhibits y-interferon-induced histocampatibility class II (Ia) antigen expression on cultured astrocytes via β2-adrenergic signal transduction mechanisms. Proc Natl Acad Sci USA 85:1292–1296

Frohman EM Frohman TC, Vayuvegula B, Gupta S, Van den Noort S (1988b) Vasoactive intestinal polypeptide inhibits the expression of the MHC class II antigens on astrocytes. J Neurol Sci 88:339–346.

Gähwiler BH (1988) Organotypic cultures of neural tissue. Trends Neurosci 11:484–489.

Goldstein GW, Betz AL (1986) The blood brain barrier. Sci Amer 255:70–79.

Gundersen K, Maehlen J (1994) Nerve-evoked electrical activity regulates molecules and cells with immunolgical function in rat muscle tissue. Eur J Neurosci 6:1113–1118

Hickey WF (1991) Migration of hematogenous cells through the blood-brain barrier and the initiation of CNS inflammation. Brain Pathol 1:97–106.

Kasid A, Morecki S, Aebersold P, Cornetta K, Culver K, Freeman S, Director E, Lotze MT, Blaese RM, Anderson WF (1990) Human gene transfer. Characterization of human infiltrating lymphocytes as vehicles for retroviral-mediated gene transfer in man. Proc Natl Acad Sci USA 87:473–477.

Kerschensteiner M, Gallmeier E, Behrens L, Klinkert WEF, Kolbeck R, Hoppe E, Stadelmann C, Lassmann H, Wekerle H, Hohlfeld R (1999) Activated human T cells, B cells and monocytes produce brain-derived neurotrophic factor (BDNF) in vitro and in brain lesions: A neuroprotective role of inflammation? J Exp Med 189:865–870.

Kiefer R, Haas CA, Kreutzberg GW (1991) Gamma interferon-like immunoreactive material in rat neurons: Evidence against a close relationship to gamma interferon. Neuroscience 45:551 – 560.

Konno H, Yamamoto T, Suzuki H, Yamamoto H, Iwasaki Y, Ohara Y, Terunuma H, Harata N (1990) Targeting of adoptively transferred experimental allergic encephalomyelitis lesion of the site of Wallerian degeneration. Acta Neuropathol 80:521–526.

Kramer R, Zhang Y, Gehrmann J, Gold R, Thoenen H, Wekerle H (1995) Gene transfer through the blood-nerve barrier: Nerve growth factor engineered neuritogenic T lymphocytes attenuate experimental autoimmune neuritis. Nature Med 1:1162–1166.

Kreutzberg GW (1996) Microglia: a sensor for pathological events in the CNS. Trends Neurosci 19:312–318.

Lee SC, Collins M, Vanguri P, Shin ML (1992) Glutamate differentially inhibits the expression of class II MHC antigens on astrocytes and microglia. J Immunol 148: 3391–3397.

Lindå H, Hammarberg H, Cullheim S, Levinovitz A, Khademi M, Olsson T (1998) Expression of MHC class I and β2-microglobulin in rat spinal motoneurons: Regulatory influences of IFN-gamma and axotomy. Exp Neurol 150:282–295.

Lo DC (1995) Neurotrophic factors and synaptic plasticity. Neuron 15:979–981.
Maehlen J, Olsson T, Zachau A, Klareskog L, Kristenssen K (1989) Local enhancement of major histocompatibility complex (MHC) class I and class II expression and cell infiltration in experimental allergic encephalomyelitis around axotomized motor neurons. J Neuroimmunol 23:125–132.
Mathisen PM, Yu M, Johnson JM, Drazba JA, Tuohy VK (1997) Treatment of experimental autoimmune encephalomyelitis with genetically modified memory T cells. J Exp Med 186:159–164.
Mavilio F, Ferrari G, Rossini S, Nobili N, Bonini C, Casorati G, Travesari C, Bordignon C (1994) Peripheral blood lymphocytes as target cells of retoviral vector-mediated gene transfer. Blood 83:1988–1997
Moritani M, Yoshimoto K, Li S, Kondo M, Iwahana H, Yamaoka T, Sano T, Nakano N, Kikutani H, Itakura M (1996) Prevention of adoptively transferred diabetes in nonobese diabetic mice with IL-10 transduced islet specific Th1 lymphocytes. A gene therapy model for autoimmune disease. J Clin Invest 98:1851–1859
Muller D, Wang C, Skibo G, Cremer H, Calaora V, Rougon G, Kiss JZ (1996) PSA-NCAM is required for activity-induced synaptic plasticity. Neuron 17:413–422
Neumann H, Wekerle H (1998) Neuronal control of the immune response in the central nervous system. Linking brain immunity to neurodegeneration. J Neuropathol Exp Neurol 58:1–9
Neumann H, Cavalié A, Jenne DE, Wekerle H (1995) Induction of MHC class I genes in neurons. Science 269:549–552
Neumann H, Boucraut J, Hahnel C, Misgeld T, Wekerle H (1996) Neuronal control of MHC class II inducibility in rat astrocytes and microglia. Eur J Neurosci 8:2582–2590
Neumann H, Schmidt H, Cavalié A, Jenne D, Wekerle H (1997a) MHC class I gene expression in single neurons of the central nervous system: Differential regulation by interferon-y and tumor necrosis factor-*α*. J Exp Med 185:305–316
Neumann H, Schmidt H, Wilharm E, Behrens L, Wekerle H (1997b) Interferon-*γ* gene expression in sensory neurons: Evidence for autocrine gene regulation. J Exp Med 186:2023–2031
Neumann H, Misgeld T, Matsumuro K, Wekerle H (1998) Neurotrophins inhibit class II inducibility of microglia: Involvement of the p75 receptor. Proc Natl Acad Sci USA 95:5779–5784
Olsson T, Kristensson K, Ljungdahl A, Maehlen J, Holmdahl R, Klareskog L (1989) Gamma-interferon-like immunoreactivity in axotomized rat motor neurons. J Neurosci 9:3870–3875
Olsson T, Diener P, Ljungdahl Å, Höjeberg B, Van der Meide P, Kristensson K (1992) Facial nerve transection causes expansion of meyelin autoreactive T cells in regional lymph nodes and T cell homing to the facial nucleus. Autoimmunology 13:117–126
Pardridge WM (1998) CNS drug design based on principles of blood-brain barrier transport. J Neurochem 70:1781–1792
Raivich G, Jones LL, Kloss CUA, Werner A, Neumann H, Kreutzberg GW (1998). Immune surveillance in the injured nervous system: T lymphocytes invade the axotomized mouse facial motor nucleus and aggregate around sites of neuronal degeneration. J Neurosci 18:5804–5816
Schnell L, Schneider R, Perry VH, Schwab ME (1997) Lymphocyte recruitment following spinal cord injury in mice is altered by prior viral exposure. Eur J Neurosci 9:1000–1007
Shaw MK, Lorens JB, Dhawan A, DalCanto R, Tse HY, Tran AB, Bonpanel C, Eswaran SL, Brocke S, Sarvetnick N, Steinman L, Nolan GP, Fathman CG (1997) Local delivery of interleukin-4 by retrovirus-transduced T lymphocytes ameliorates experimental autoimmune encephalomyelitis. J Exp Med 185:1711–1714
Tchélingérian J-L, Quinonero J, Booss J, Jacques C (1993) Localization of TNF-*α* and IL-1*α* immunoreactivities in striatal neurons after surgical injury to the hippocampus. Neuron 10:213–224
Thoenen H (1995) Neurotrophins and neuronal plasticity. Science 270:593–598
Torcia M, Bracci-Laudiero L, Lucibello M, Nencioni L, Labardi D, Rubatelli A, Cozzolino F, Aloe L, Garaci E (1996) Nerve growth factor is an autocrine survival factor for memory B lymphocytes Cell 85:345–356
Verma IM, Somia N (1997) Gene therapy – promises, problems and prospects. Nature 389:239–242
Woffendin C, Yang ZY, Udaykumar, Xu L, Yang NS, Sheehy MJ, Nabel G (1994) Nonviral and viral delivery of a human immunodeficiency virus protective gene into primary human T cells. Proc Natl Acad Sci USA 91:11581–11585

Regulation of Cyclic AMP Synthesis in Microglial Cells and Possible Role of Cyclic AMP in Neuroprotection

G. Levi

Summary

Microglial cells are believed to play an active role in most inflammatory and degenerative brain diseases. Many of the effects of microglial cells can be ascribed to the numerous substances that these cells can synthesize and release upon activation by a variety of stimuli, including cytokines, pro-inflammatory substances, neurotransmitters and viral proteins.

Cyclic AMP (cAMP) is one of the major second messengers and is known to regulate several properties of activated microglia. Interestingly, elevations of cAMP levels can elicit neuroprotective processes in vivo and in vitro, which could be in part related to a controlled microglial activation. It is therefore important on one hand to understand the factors regulating the synthesis of cAMP, and on the other hand to elucidate how cAMP can regulate microglial functions related to brain inflammation, degeneration and repair. In cultured microglia, at least three types of receptors linked to adenylyl cyclase through a stimulatory G protein have been described: the β-adrenergic, the adenosine A_{2a} and the prostaglandin (PG) EP2 receptors. The microglial synthesis of cAMP can be depressed by a variety of pathophysiological stimuli that can act at different levels in the cAMP cascade. For example, bacterial lipopolysaccharide and interferon-γ have been shown to lower cAMP levels by stimulating type IV phosphodiesterase activity. Acute stimulation of protein kinase C led to a depression of PGE_2-induced cAMP synthesis by interfering with EP2/EP4 receptors, whereas inactivation or inhibition of protein kinase C reduced the synthesis of the second messenger by inhibiting adenylyl cyclase activity. The human immunodeficiency virus protein gp 120 inhibited β-adrenergic-mediated cAMP synthesis but not the synthesis induced by direct activation of adenylyl cyclase, whereas the viral regulatory protein *tat* inhibited cAMP synthesis in both conditions. In lipopolysaccharide-activated microglial cells, the effects of cAMP were tested by different authors using one or more of the following tools: receptor agonists (isoproterenol, PGE_2, 11-deoxy-16,16-dm-PGE_2, adenosine agonists), adenylyl cyclase activators or inhibitors, inhibitors of protein kinase A, cAMP analogs. cAMP analogs and cAMP-elevating agents (including PGE_2) depressed the production of substances and/or expression of molecules potentially involved in the establishment of brain damage (nitrogen and oxygen radicals) or in the amplification of inflammation and immune responses (tumor necrosis

Patterson/Kordon/Christen (Eds.)
Neuro-Immune Interactions in Neurologic and Psychiatric Disorders

factor, interleukin-1, interleukin-12 and co-stimulatory molecules) while enhancing that of potentially neuroprotectivc or immunodepressive substances (interleukin-10, PGE_2).

Introduction

Although the role of microglial cells (the resident brain macrophages) in the normal brain is largely unknown, many studies published in the last five to 10 years deal with the function of this relatively abundant cell population (10–20% of total glial cells) in pathological conditions (see reviews by Banati and Graeber 1994; Mallat and Chamak 1994; Perry 1994; McGeer and McGeer l995; Kreutzberg 1996; Moore and Thanos 1996; Minghetti and Levi 1998). It is now widely accepted that in the normal adult brain microglial cells are kept in a quiescent, resting state by a set of local neuronal and astroglial interactions involving signals whose nature must still be unraveled. However, microglial cells maintain a high degree of reactivity, which makes them capable of being "activated" by a variety of pathological stimuli, enabling them to have a prominent role in host defense on one side and in the establishment and maintenance of brain damage on the other. The balance between defensive and repair reactions and aggressive, pro-inflammatory and neurotoxic reactions will determine the final outcome of pathological processes affecting the central nervous system. Table 1 provides a list of conditions that can induce microglial cells to produce a large number of substances that may be either neurotrophic or neurotoxic, depending not only on the nature of the particular molecule produced or expressed but also on the magnitude and duration of the microglial response.

Table 1. Products of activated microglia and inducers of microglia activation [a]

Products	
Growth factors	NGF, bFGF, TGFs, HGF, others
Cytokines	Interleukins 1, 3, 5, 6, 12; TNF; MCP-1; MIP-1
Coagulation factors	Plasminogen, Plasminogen activator
Free radicals	Superoxide, NO
Lipid mediators	PGD_2, PGE_2, TXA_2, LTB_4
Enzymes	PLA-2, COX-2, iNOS, metalloprotease inhibitors, elastase, gelatinase
Neurotoxins	β-amyloid, quinolinate, glutamate, others
Complement	C1, C3, C4
Extracellular matrix	Thrombospondin
Receptors	Purinergic, MHC-I, MHC-II, LCA, LFA-1, C3 and C4, Fc, thrombin, CSF-1, vitronectin, cytokines, trk-B
Inducers	
Trauma, demyelination, excitotoxins, degeneration, virus infection, ischemia, hormones, neurotransmitters, phorbol esters, LPS, inflammatory cytokines, irradiation, ionophores, β-amyloid, others	

[a] For references, see McGeer and McGeer 1995; Minghetti and Levi 1998.

Table 2. Neuroprotective action of cAMP-elevating agents

Agent	Protective effect
Rolipram	MPTP toxicity in vivo [a] EAE [b] Neuronal ischemic damage [c]
Forskolin	PC12 cell apoptosis induced by H_2O_2 [d]
Isoproterenol	EAE [e] Microglial neurotoxicity in vitro [f]
Prostaglandins	NMDA toxicity in vitro [g] Hypoxia/glutamate toxicity in vitro [h] Microglial neurotoxicity in vitro [f] EAE [i]
Adenosine-related drugs	Glutamate toxicity [j] Ischemic damage [k]

[a] Hulley et al. (1995)
[b] Genain et al. (1995); Jung et al. (1996); Sommer et al. (1995)
[c] Kato et al. (1995)
[d] Kamata et al. (1996)
[e] Wiegmann et al. (1995)
[f] Théry et al. (1994)
[g] Akaike et al. (1994)
[h] Cazevieille et al. (1993, 1994); Otsuki et al. (1994)
[i] Reder et al. (1994)
[j] Finn et al. (1991)
[k] Schubert et al. (1997).

These introductory statements underline the importance of understanding what modulates the release of these compounds, not only for the comprehension of disease pathogenesis but also to provide clues for the pharmacological orientation of microglia towards neuroprotection. Evidence is accumulating that the second messenger cyclic AMP (cAMP) is an important regulator of the production of several substances by activated microglial cells, as I will briefly illustrate below. Interestingly, it has been reported that cAMP (or cAMP elevating agents) exerts a neuroprotective role in several in vivo and in vitro models of brain pathology (see the examples provided in Table 2). Thus, one can speculate that cAMP may be neuroprotective through its regulatory action on microglia. It is therefore essential to know what regulates cAMP synthesis and what microglial functions are under the control of cAMP. This is the topic of the present article, which will deal with studies performed in our and other laboratories with purified microglial cultures obtained from the neonatal rodent brain.

Regulation of the cAMP System in Microglia

Regulation of the cAMP system can be exerted at any of the steps in the cAMP cascade (receptors, Gi or Gs proteins, adenylyl cyclase, phosphodiesterases and protein kinase A). In the case of microglial cells, at least three classes of compounds acting at the level of three different types of Gs protein-linked receptors have been shown to elicit large increases in the synthesis of the cyclic nucleotide: β-adrenergic agonists, such as isoproterenol (Levi et al. 1993; Patrizio et al. 1995, 1996), prostaglandin E_2 (PGE_2) receptor agonists, such as PGE_2 itself or its EP2 receptor selective agonist 11-deoxy-16,16-dm-PGE_2 (Patrizio et al. 1995, 1996 and unpublished results), and adenosine A_{2a} receptor agonists (Fiebich et al 1996).

Pathological signals can affect cAMP responses. An example gives me the opportunity to show a regulation of microglial cAMP concentrations at the level of phosphodiesterase, the enzyme responsible for the degradation of the cyclic nucleotide. We have observed that a short (10 min) treatment of microglial cultures with either of two classical pro-inflammatory, macrophage-activating substances, bacterial lipopolysaccharide (LPS) and the cytokine interferon-γ (IFN-γ), caused a substantial reduction in the subsequent cAMP accumulation induced by two agonists acting at Gs-coupled receptors (isoproterenol and PGE_2) or by the adenylyl cyclase activator forskolin. The level of action of the two pro-inflammatory substances could not be the receptors or G proteins, which were bypassed in the forskolin experiments, neither could it be adenylyl cyclase, since a preincubation period was required to observe the inhibitory effect. The observation that specific inhibitors of type IV phosphodiesterase (rolipram or Ro20-1788), but not inhibitors of other isoforms of the enzyme, counteracted the effects of LPS and IFN-γ supports the idea that these two substances, though acting at different receptors, suppressed cAMP accumulation through a common mechanism, namely acceleration of its metabolic breakdown (Patrizio et al. 1995).

The cAMP system can be affected not only by exogenous substances that interfere at one or the other step of the cAMP cascade but also by interaction with other transduction systems. We investigated the interaction with the protein kinase C (PKC) system (Patrizio et al. 1997). The rational and the background for these studies, which were performed using both astrocyte and microglial cultures, are outlined in Table 3. A different regulation of the cAMP system by PKC in astrocytes and microglia might help to explain qual-

Table 3. Regulation of cAMP by PKC (background)

- Several (astro-micro)-glial activators (LPS, IFN-γ, TNF, TGF-β, IL-10) utilize signal transduction pathways involving PKC
- PKC may activate or depress the cAMP system depending on the cell type
- PKC may act at different levels of the cAMP signaling system: receptors, G proteins, adenylyl cyclase, phosphodiesterases.

itative and quantitative differences in the response of these cells to various physiological and pathological stimuli.

Experiments with astrocyte or microglia cultures showed different, almost opposite regulations of cAMP responses: in astrocytes, a 10-min treatment with the phorbol esters phorbol 12-myristate 13-acetate (PMA) and 4β-phorbol-12-13 didecanoate (4β-PDD; but not with the inactive isomer 4α-PDD), or with diacylglycerol, all of which activate PKC, caused a dose-dependent increase in the accumulation of cAMP induced by isoproterenol and forskolin. The PMA-induced enhancement was prevented by the PKC inhibitors staurosporine and calphostin-C, which, by themselves, were devoid of effect, and by a more prolonged treatment with PMA (100 nM) that down-regulates PKC. In microglia, a 10-min treatment with PMA or PKC inhibitors did not alter cAMP accumulation, whereas longer treatments with PMA or 4β-PDD (but not 4α-PDD) led to a time- and dose-dependent depression of the cAMP response. The inhibitory effect of phorbol esters was mimicked, rather than prevented, by staurosporine and calphostin-C. Several observations indicated that the effects of PKC activators on the astroglial cAMP system were not related to an interaction with membrane receptors (that were bypassed by forskolin), or with G proteins or phosphodiesterases, since the effect was maintained in cells treated with pertussis toxin or with a non-specific phosphodiesterase inhibitor. Instead, the target of PKC appeared to be adenylyl cyclase, whose activity was doubled in astrocytes treated for 10 min with PMA. Conversely, adenylyl cyclase was substantially inhibited in microglial cells treated with PMA or calphostin C for 4 hours (Patrizio et al. 1997).

Although forskolin- and isoproterenol-induced cAMP generation in microglia was not affected by PKC activation, PGE_2-induced cAMP accumulation was strongly depressed by PMA in a time-dependent (Fig. 1) and dose-dependent way (not shown; maximum inhibition with 100 nM PMA).

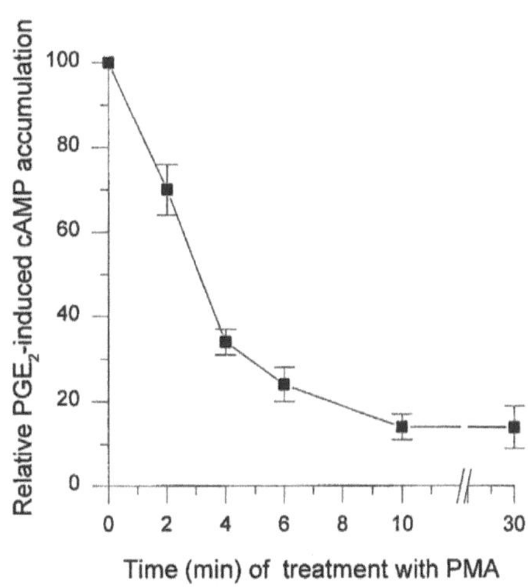

Fig. 1. Time-dependence of the inhibitory effect of PMA on PGE_2-induced cAMP accumulation in microglia. Rat microglial cultures were treated for different times with 100 nM PMA and then exposed for 10 min to 1µM PGE_2. The concentration of cAMP in control cultures (treated with PGE_2, but not with PMA) was 2.5 pmol/µg protein. This value was taken as 100. Means ± SEM of three experiments are presented (Patrizio and Levi, unpublished data)

Table 4. Acute PMA pretreatment decreases cAMP accumulation induced by the selective EP2/EP4 agonist, 11-deoxy-16,16-dm-PGE_2 (dm-PGE_2)[a]

	cAMP concentration (pmol/100 μg protein)		
		100 nM PMA	% inhibition
dm-PGE_2 100 nM	3 ± 0.5	1.7 ± 1	43
dm-PGE_2 1 μM	5 ± 0.7	2.8 ± 1	44
dm-PGE_2 10 μM	22 ± 3	9 ± 3	59
dm-PGE_2 100 μM	193 ± 50	30 ± 12	84

[a] Microglial cultures were exposed to 100 nM PMA for 10 min, then the cells were stimulated with increasing concentrations of dm-PGE_2 for 10 min and assayed for cAMP content and protein. Means ± SEM of three independent experiments run in duplicate (Patrizio and Levi, unpublished data).

Table 5. PKC effect on the cAMP system in glial cells[a]

	cAMP response	
	Astrocytes	Microglia
PKC activation (10 min PMA)		
Isoproterenol	↑	no effect
Forskolin	↑	no effect
PGE_2	no cAMP response	↓
Target	*Adenylyl cyclase (stimulation)*	*EP2/EP4 receptors*
PKC down-regulation/inhibition (4h PMA, PKC inhibitors)		
Isoproterenol	no effect	↓
Forskolin	no effect	↓
Target		*Adenylyl cyclase (inhibition)*

[a] PKC activation potentiates isoproterenol and forskolin-induced cAMP synthesis in astrocytes (but not in microglia) by activating adenylyl cyclase, and inhibits PGE_2-induced cAMP synthesis in microglia by interacting with EP2 receptors. PKC down-regulation or inhibition does not affect cAMP responses in astrocytes and inhibits isoproterenol- and forskolin-induced cAMP synthesis in microglia by depressing the cyclase activity. For other details, see text.

The lack of effect of PMA on isoproterenol and forskolin excludes that the effect on PGE_2 is related to PKC interaction with G proteins, adenylyl cyclase and phosphodiesterases. The remaining possible target is the EP2 or EP4 receptor, as supported by the observation that the inhibitory effect of PMA was also seen when cAMP synthesis was stimulated with the specific agonist 11-deoxy-16,16-dm-PGE_2 (Table 4).

Table 5 summarizes the interactions between PKC and the cAMP system described above. The reasons for the nearly opposite behavior of astrocytes and microglia still have to be elucidated. It is legitimate to think that they are related to the different expression of PKC or adenylyl cyclase isoforms in the two cell types. Indeed, preliminary Western blot analysis observations provided evidence for a different pattern of expression of PKC isoforms in astrocytes and microglia. In particular, at least in our experimental conditions, microglial cells did not express detectable amounts of PKC isoforms *α* and *ε* (Slepko et al. 1999), which may suggest that the activation of one or both these isoforms is responsible for the phosphorylation and activation of adenylyl cyclase in astrocytes.

As already mentioned in the case of the inhibitory effects of LPS and IFN-*γ*, cAMP accumulation in glial cells can be affected by substances present in the brain in pathological conditions. I would like to add another couple of examples, although in these cases the level of action of the pathological stimulus in the cAMP cascade is still unknown. A few years ago we reported that acute exposure to the human immunodeficiency virus (HIV) envelope protein gp120 inhibits the accumulation of cAMP induced by isoproterenol, but not that induced by forskolin, in rat astrocyte and microglial cultures and in a human astrocytoma cell line (Levi et al. 1993). Such inhibition was found to have functional consequences, both in astrocytes which showed a decreased phosphorylation of glial fibrillary acidic protein in response to isoproterenol, and in microglia, which showed a decrease in the isoproterenol-induced inhibition of LPS-evoked tumor necrosis factor-*α* (TNF-*α*) production. Experiments currently in progress in our laboratory indicate that another HIV protein, the regulatory and transactivating protein *tat*, causes a time-dependent inhibition of isoproterenol-induced cAMP synthesis in microglia (Fig. 2).

All four factors mentioned above that decrease cAMP responses have been implicated in the establishment of tissue damage in the central nervous system. To better understand whether decreased cAMP responses might be

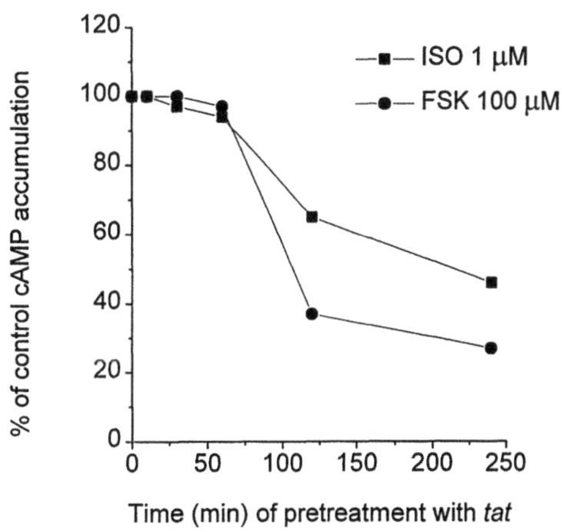

Fig. 2. Time-dependence of the inhibitory effect of *tat* on cAMP accumulation in microglia. Rat microglial cultures were treated with *tat* protein (100 ng/ml) for the time periods indicated. Then they were exposed to 1 μM isoproterenol (ISO) or 100 μM forskolin (FSK) for 10 min and assayed for cAMP content and protein. The accumulation of cAMP induced in the absence of *tat* was taken as 100. Data from one representative experiment (Patrizio and Levi, unpublished data)

involved in neuropathogenesis, it seemed important to assess what microglial functions are under the control of cAMP. I will briefly describe experiments along these lines performed by us and by others.

Regulation by cAMP of Microglial Functions Related to Inflammation, Neurotoxicity and Immune Responses

Regulation of Prostanoid and Nitric Oxide (NO) Synthesis

There are several reasons that prompted us to study the regulation of prostanoid (that include prostaglandins, prostacyclin and thromboxane) and NO metabolic pathways. In inflammatory cells, including microglia, the production of prostanoids and NO is largely controlled by the expression of inducible rate-limiting enzymes, cyclooxygenase-2 (COX-2) and inducible nitric oxide synthase (iNOS or NOS-II), respectively, whereas the corresponding constitutive enzymes (COX-1 and cNOS) are scarcely, if at all, expressed. The stimuli inducing the expression of these enzymes are often, even if not always, similar (LPS, cytokines). Both inducible enzymes are codified by immediate early genes and can lead to the production of large amounts of end-products. Such end-products, in turn, control, often in an opposite way, similar functions, such as vascular tone, inflammation, immune responses and neurotransmitter release, and mutually regulate the expression of the rate-limiting inducible enzymes (reviewed by Minghetti and Levi 1998). Last but not least, our interest was stimulated by recent evidence provided in the literature in favor of a neuroprotective role of some prostaglandins (notably,

Table 6. Possible beneficial and detrimental activities of prostanoids and NO in the CNS[a]

Prostanoids Possible beneficial roles	NO and/or NO-derived free radicals Proinflammatory and toxic effects
Anti-inflammatory activities Elevation of cAMP in cells expressing specific receptors	Pro-inflammatory activity Induction of proinflammatory cytokines Increase in vascular permeability
Neuroprotective roles Protection of neuronal cultures from: macrophage toxicity hypoxia glutamate toxicity	Cytotoxic activities Induction of cell death in many cell types, including neurons and oligodendrocytes in CNS
Protective effect in EAE Increased production of PGE_2 concomitant with recovery phase of EAE Improvement of EAE with stable PGE_2 analogs	Phatogenetic role in autoimmune and neurodegenerative diseases Suggested by the increase in iNOS and by the protective roles of iNOS inhibitors in various diseases including EAE

[a] For references, see text and Minghetti and Levi 1998.

PGE_2 and PGI_2), as opposed to the well-accepted neurotoxic activity of excessive NO and/or NO-derived nitrogen radicals (see Table 6).

Thus, the balance between prostanoid and NO production in activated microglia may be an important factor in determining whether the microglial response to a given stimulus or set of stimuli will result in neuroprotection or cell damage.

Several lines of evidence indicate that cAMP potentiates the expression of COX-2 when induced by a typical microglial activator such as LPS (Fiebich et al. 1996; Minghetti et al. 1997a), and at the same time depresses the expression of iNOS (Minghetti et al. 1997b). Support for this contention was provided by experiments with cAMP analogs (dibutyryl cAMP, or the thiolated Sp-cAMP isomer), adenylyl cyclase activators (forskolin), and cAMP elevators acting on Gs protein-linked receptors (isoproterenol, PGE_2, adenosine agonists), as well as by experiments in which either cAMP synthesis was prevented by inhibiting adenylyl cyclase with the specific inhibitor SQ22,536 (9-tetrahydro-2-furanyl-9H-purine-6-amine) or cAMP activity was prevented by a competitive inhibitor acting at the level of protein kinase A (thiolated inactive Rp-cAMP isomer). Changes in the expression of COX-2 and iNOS were always accompanied by parallel changes in the production of PGE_2 and NO (measured as nitrite), respectively.

On the basis of the above observations, one can envisage the existence of a cAMP-PGE_2-NO loop, whereby cAMP enhances PGE_2 and depresses NO production. The decreased NO production will alleviate the inhibitory effect of NO on COX-2 expression (Minghetti et al. 1996)whereas the increased release of PGE_2 will further feed the circuit by stimulating the synthesis of the second messenger, thus maintaining a high level of COX-2 (an enzyme with short half-life) and a low level of iNOS. The neuromodulator adenosine may also impinge upon the loop via its A_{2a} receptors in pathological conditions

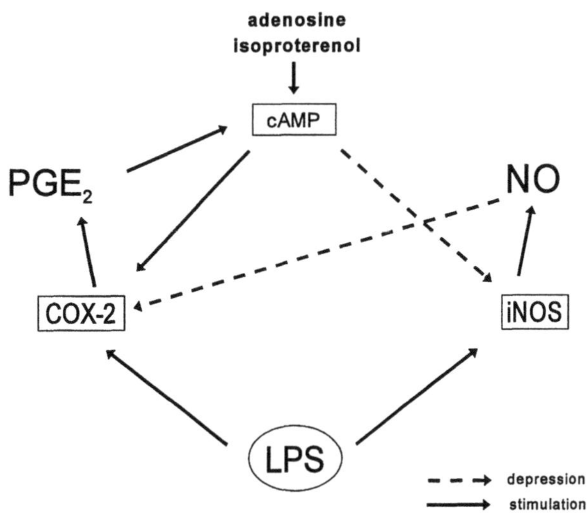

Fig. 3. The PGE_2-cAMP-NO loop. See text for explanation

(trauma, ischemia) in which it is released in large quantities in brain tissue (Martin et al. 1994). Another feeder of the loop may be noradrenaline, through its microglial β-adrenergic receptors (Fig. 3). Interestingly, cAMP seems to be able to depress not only the production of nitrogen radicals, but also that of oxygen radicals, as suggested by the inhibitory effect of isoproterenol on the generation of superoxide in hamster microglial cultures (Colton and Chernyshev 1996).

One can wonder what may stop the apparently self-potentiating PGE_2-cAMP-NO loop described above. One possibility is that other factors, not considered in the above paragraph, come into play. For example, in a previous section, several "pathological" substances (LPS, IFN-γ, HIV proteins) have been mentioned that depress cAMP responses. Of these, IFN-γ has been also shown to depress the expression of COX-2 and the production of PGE_2 (Minghetti et al. 1996). Moreover, interleukin-10 (IL-10), whose release from macrophages (Strassmann et al. 1994) and microglia (see below) is enhanced by PGE_2, was recently reported to inhibit the expression of COX-2 and the production of PGE_2 in LPS-treated rat microglial cultures (Minghetti et al. 1998). Such feed-back regulation may prevent excessive depression of microglial reactivity by IL10, PGE_2 and cAMP. Saturation of membrane receptors by the active exogenous ligands and/or of protein kinase A by cAMP may be other mechanisms limiting the self-potentiating activity of the loop.

Regulation of Microglial Immune Functions

Cyclic AMP and PGE_2 (through cAMP) appear to exert a wide control of microglial functions related to the role of these cells in local immune responses. One aspect concerns the regulation of the expression of B7 antigens. In order to act as an antigen-presenting cell, a cell must be able to express major histocompatibility complex (MHC) antigens and a set of so called co-stimulatory molecules. Among these, B7 molecules (B7-1 and B7-2) are expressed on professional antigen-presenting cells (dendritic cells, B-cells, macrophages) and, after interacting with specific receptors on T cells (CD28, CTLA4), function as powerful co-stimulators of T cell proliferation and cytokine production. Blockade of the interaction between B7 and its ligand CD28 was shown to prevent or ameliorate autoimmune diseases (Khoury et al. 1995; Lenschow et al. 1996). Microglia can be induced to express B7 molecules, and regulation of their expression may affect the antigen-presenting ability of these cells. It is interesting that PGE_2, isoproterenol and dibutyryl cAMP were able to inhibit the expression of B7-2 in mouse microglia stimulated with LPS and IFN-γ (Menèndez Iglesias et a1. 1997).

The microglial expression and secretion of several cytokines involved in inflammatory and immune responses were also affected by cAMP analogs and cAMP elevating agents, including PGE_2, according to a characteristic pattern. We shall consider here the effects on the production of IL-10, TNF-α, IL-12 and IL-1. IL-10 is an important macrophage deactivator, produced by the Th2 subtype of T cells, B-cells and macrophages, including microglia, and is a potent inhibitor of inflammatory and Th1-cell mediated immune responses (Moore et al. 1993). TNF-α is a pro-inflammatory cytokine that, alone or in concert with other cytokines, has been considered to be responsible for cellular damage or death in the central nervous system (Merrill 1992).

IL-12, a 75 kDa heterodimeric protein produced by dendritic cells and monocytes/macrophages, including microglia (see Aloisi et al. 1997 also for other references), is a potent inducer of IFN-γ production by T and NK cells and enhances proliferation and cytolytic activity of these cells. In particular, it promotes the development of Th1 cells and has been implicated in the induction of Th1-mediated, organ-specific, autoimmune diseases (Trembleau et al. 1995), including experimental autoimmune encephalomyelitis (EAE, Leonard et al. 1995) and multiple sclerosis (Windhagen et al. 1995). IL-1 is a pro-inflammatory cytokine, mainly produced by monocytes/macrophages and microglia, and its cerebral concentration is increased in several cerebral diseases, including Alzheimer's and inflammatory demyelinating diseases, in which a pathogenetic role has been proposed (Merrill 1992). In mouse microglial cultures stimulated with LPS+IFN-γ, cAMP-elevating agents, including PGE_2, enhanced the expression and secretion of IL-10, depressed that of TNF-α (Aloisi et al. 1999) and down-regulated that of bioactive IL-12 protein p75 and of the p40 chain (Aloisi et al. 1997). The LPS-induced production of IL-1 was found to be depressed by isoproterenol in microglial cultures obtained from the mouse brain (Hetier et al. 1991), where, however, the decreased release of the cytokine was not paralleled by a decreased expression of IL-1 mRNA. In another more recent study, Caggiano and Kraig (1998) reported that IL-1 production was inhibited by PGE_2 in LPS-activated rat microglial cultures. In the latter case the involvement of cAMP has not been considered as a possible mechanism, and the authors favored the idea that the inhibition of IL-1 release was related to the ability of PGE_2 to prevent the expression of outward rectifying potassium currents, which are typical of activated microglia (Kettenmann et al. 1990; Nörenberg et al. 1992; Visentin et al. 1995). Although this interpretation was supported by the observation that a classical inhibitor of such currents, 4-amino pyridine, prevented IL-1 release in the same way as PGE_2, and that the two substances did not exhibit an additive effect, the involvement of cAMP in the PGE_2-mediated effects remains a possibility that cannot be ruled out.

Conclusions

Of all functional parameters examined, cAMP analogs and cAMP-elevating agents always depressed those related to neurotoxicity, inflammation and immune reactions (see the effects on NO, superoxide, B7 molecules, TNF-α, IL-1, IL-12) and enhanced those related to macrophage deactivation and immunosuppression (such as IL-10 and PGE_2 production). Altogether, we can predict that in most circumstances elevations of cAMP in microglia will cause a reduction of inflammation, immune responses and tissue damage in the central nervous system.

Acknowledgments. This work was supported by the following grants: Project on AIDS of the Italian Ministry of Health; Project on Multiple Sclerosis of the Istituto Superiore di Sanità.

References

Akaike A, Kaneko S, Tamura Y, Nakata N, Shiomi H, Ushikubi F, Narumiya S (1994) Prostaglandin E_2 protects cultured cortical neurons against *N*-methyl-D-aspartate receptor-mediated glutamate cytotoxicity. Brain Res 663:237–243

Aloisi F, Penna G, Cerase J, Menèndez Iglesias B, Adorini L (1997) IL-12 production by central nervous system microglia is inhibited by astrocytes. J Immunol 159:1604–1612

Aloisi F, De Simone R, Columba-Cabezas S, Levi G (1999) Opposite effects of interferon-γ and prostaglandin E_2 on tumor necrosis factor and interleukin-10 production in microglia: a regulatory loop controlling microglia pro- and anti-inflammatory activies. J Neurosci Res 56:571–580

Banati RB, Graeber MB (1994) Surveillance, intervention and cytotoxicity: Is there a protective role of microglia. Dev Neurosci 16:114–127

Caggiano AO, Kraig RP (1998) Prostaglandin E_2 and 4-aminopyridine prevent the lipopolysaccharide-induced outwardly rectifying potassium current and interleukin-1β production in cultured rat microglia. J Neurochem 70:2357–2368

Cazevieille C, Muller A, Bonne C (1993) Prostacyclin (PGI_2) protects rat cortical neurons in culture against hypoxia/reoxygenation and glutamate-induced injury. Neurosci Lett 160:106–108

Cazevieille C, Muller A, Meynier F, Dutrait N, Bonne C (1994) Protection by prostaglandins from glutamate toxicity in cortical neurons. Neurochem Int 24:395–398

Colton CA, Chernyshev ON (1996) Inhibition of microglial superoxide anion production by isoproterenol and dexamethasone. Neurochem Int 29:43–53

Fiebich BL, Biber K, Lieb K, Van Calker D, Berger M, Bauer J, Gebicke-Haerter PJ (1996) Cyclooxygenase-2 expression in rat microglia is induced by adenosine A_{2a}-receptors. Glia 18:152–160

Finn SF, Swartz KJ, Beal MF (1991) 2-Chloroadenosine attenuates NMDA, kainate, and quisqualate toxicity. Neurosci Lett 126:191–194

Genain CP, Roberts T, Davis RL, Nguyen M-H, Uccelli A, Faulds D, Li Y, Hedgpeth J, Hauser SL (1995) Prevention of autoimmune demyelination in non-human primates by a cAMP-specific phosphodiesterase inhibitor. Proc Natl Acad Sci USA 92:3601–3605

Hetier E, Ayala J, Bousseau A, Prochiantz A (1991) Modulation of interleukin-1 and tumor necrosis factor expression by β-adrenergic agonists in mouse ameboid microglial cells. Exp Brain Res 86:407–413

Hulley P, Hartikka J, Abdel'Al S, Engels P, Buerki HR, Wiederhold KH, Müller T, Kelly P, Lowe D, Lübbert H (1995) Inhibitors of type IV phosphodiesterases reduce the toxicity of MPTP in substantia nigra neurons *in vivo.* Eur J Neurosci 7:2431–2440

Jung S, Zielasek J, Köllner G, Donhauser T, Toyka K, Hartung HP (1996) Preventive but not therapeutic application of Rolipram ameliorates experimental autoimmune encephalomyelitis in Lewis rats. J Neuroimmunol 68:1–11

Kamata H, Tanaka C, Yagisawa H, Hirata H (1996) Nerve growth factor and forskolin prevent H_2O_2-induced apoptosis in PC12 cells by glutathione independent mechanism. Neurosci Lett 212:179–182

Kato H, Araki T, Itoyama Y, Kogure K (1995) Rolipram, a cyclic AMP-selective phosphodiesterase inhibitor, reduces neuronal damage following cerebral ischemia in the gerbil. Eur J Pharmacol 272:107–110

Kettenmann H, Hoppe D, Gottmann K, Banati R, Kreutzberg G (1990) Cultured microglial cells have a distinct pattern of membrane channels different from peritoneal macrophages. J Neurosci Res 26:278–287

Khoury SJ, Akalin E, Chandraker A, Turka LA, Linsley PS, Sayegh MH, Hancock WW (1995) CD28-B7 costimulatory blockade by CTLA4Ig prevents actively induced experimental autoimmune encephalomyelitis and inhibits Th1 but spares Th2 cytokines in the central nervous system. J Immunol 155:4521–4524

Kreutzberg GW (1996) Microglia: a sensor for pathological events in the CNS. Trends Neurosci 19:312–318

Lenschow DJ, Walunas TL, Bluestone JA (1996) CD28/B7 system of T cell costimulation. Annu Rev Immunol 14:233–258

Leonard JP, Waldburger KE, Goldman SJ (1995) Prevention of experimental autoimmune encephalomyelitis by antibodies against interleukin 12. J Exp Med 181:381–386

Levi G, Patrizio M, Bernardo A, Petrucci TC, Agresti C (1993) Human immunodeficiency virus coat protein gp120 inhibits the β-adrenergic regulation of astroglial and microglial functions. Proc Natl Acad Sci USA 90:1541–1545

Mallat M, Chamak B (1994) Brain macrophages: Neurotoxic or neurotrophic effector cells. J Leukocyte Biol 56:416–422

Martin RL, Lloyd HGE, Cowan AI (1994) The early events of oxygen and glucose deprivation: Setting the scene for neuronal death. Trends Neurosci 17:251–257

McGeer PL, McGeer EG (1995) The inflammatory response system of brain: Implications for therapy of Alzheimer and other neurodegenerative diseases. Brain Res Rev 21:195–218

Menèndez Iglesias B, Cerase J, Ceracchini C, Levi G, Aloisi F (1997) Analysis of B7-1 and B7-2 costimulatory ligands in cultured mouse microglia: upregulation by interferon-γ and lipopolysaccharide and downregulation by interleukin-10, prostaglandin E_2 and cyclic AMP-elevating agents. J Neuroimmunol 72:83–93

Merrill JE (1992) Tumor necrosis factor alpha, interleukin-l and related cytokines in brain development: normal and pathological. Dev Neurosci 14:1–10

Minghetti L, Levi G (1998) Microglia as effector cells in brain damage and repair: focus on prostanoids and nitric oxide. Prog Neurobiol 54:99–125

Minghetti L, Polazzi E, Nicolini A, Créminon C, Levi G (1996) Interferon-γ and nitric oxide down-regulate lipopolysaccharide-induced prostanoid production in cultured rat microglial cells by inhibiting cyclooxygenase-2 expression. J Neurochem 66:1963–1970

Minghetti L, Polazzi E, Nicolini A, Créminon C, Levi G (1997a) Upregulation of cyclooxygenase-2 expression in cultured microglia by prostaglandin E_2, cyclic AMP and non steroidal anti-inflammatory drugs. Eur J Neurosci 9:934–940

Minghetti L, Nicolini A, Polazzi E, Créminon C, Maclouf J, Levi G (1997b) Inducible nitric oxide synthase expression in activated rat microglia cultures is downregulated by exogenouis prostaglandin E_2 and by cyclooxygenase inhibitors. Glia 19:152–160

Minghetti L, Polazzi E, Nicolini A, Levi G (1998) Opposite regulation of prostaglandin E_2 synthesis by transforming growth factor-β1 and interleukin 10 in activated microglia cultures. J Neuroimmunol 82:31–39

Moore S, Thanos S (1996) The concept of microglia in relation to central nervous system disease and regeneration. Prog Neurobiol 48:441–460

Moore KW, O'Garra A, De Waal Malefyt R, Vieira P, Mosmann TR (1993) Interleukin-10. Annu Rev Immunol 11:165–190

Nörenberg W, Gebicke-Haerter PJ, Illes P (1992) Inflammatory stimuli induce a new K+ outward current in cultured rat microglia. Neurosci Lett 147:171–174

Otsuki H, Yamada K, Yuguchi T, Taneda M, Hayakawa T (1994) Prostaglandin E_1 induces c-Fos and Myc proteins and protects rat hippocampal cells against hypoxic injury. J Cereb Blood Flow Metab 14:150–155

Patrizio M, Costa T, Levi G (1995) Interferon-γ and lipopolisaccharide reduce cAMP responses in cultured glial cells: reversal by a type IV phosphodiesterase inhibitor. Glia 14:94–100

Patrizio M, Riitano D, Costa T, Levi G (1996) Selective enhancement by serum factors of cyclic AMP accumulation in microglia cultures. Neurochem Int 29:89–96

Patrizio M, Slepko N, Levi G (1997) Opposite regulation of adenylyl cyclase by protein kinase C in astrocyte and microglia cultures. J Neurochem 69:1267–1277

Perry VH (1994) Macrophages and the nervous system. R.G.Landes Company, CRE Press, Austin, Texas

Reder AT, Thapar M, Sapugay AM, Jensen MA (1994) Prostaglandins and inhibitors of arachidonate metabolism suppress experimental allergic encephalomyelitis. J Neuroimmunol 54:117–127

Schubert P, Ogata T, Marchini C, Ferroni S, Rudolphi K (1997) Protective mechanisms of adenosine in neurons and glial cells. Ann NY Acad Sci 825:1–10

Slepko N, Patrizio M, Levi G (1999) Expression and translocation of protein kinase C isoforms in rat microglial and astroglial cultures. J Neurosci Res, in press

Sommer N, Löschmann P-A, Northoff GH, Weller M, Steinbrecher A, Steinbach JP, Lichtenfels R, Meyermann R, Riethmüller A, Fontana A, Dichgans J, Martin R (1995) The antidepressant rolipram suppresses cytokine production and prevents autoimmune encephalomyelitis. Nature Med 1:244–248

Strassmann G, Patil-Koota V, Finkelman F, Fong M, Kambayashi T (1994) Evidence for the involvement of interleukin-10 in the differential deactivation of murine peritoneal macrophages by prostaglandin E_2. J Exp Med 180:2365–2370

Théry C, Dobbertin A, Mallat M (1994) Downregulation of in vitro neurotoxicity of brain macrophages by prostaglandin E_2 and a β-adrenergic agonist. Glia 11:383–386

Trembleau S, Germann T, Gately MK, Adorini L (1995) The role of IL-12 in the induction of organ-specific autoimmune diseases. Immunol Today 16:383–386

Visentin S, Agresti C, Patrizio M, Levi G (1995) Ion channels in rat microglia and their different sensitivity to lipopolysaccharide and interferon-γ. J Neurosci Res 42:439–451

Wiegmann K, Muthyala S, Kim DH, Arnason BGW, Chelmicka-Schorr E (1995) β-Adrenergic agonists suppress chronic/relapsing experimental allergic encephalomyelitis (CREAE) in Lewis rats. J Neuroimmunol 56:201–206

Windhagen A, Newcombe J, Dangond F, Strand C, Woodroofe MN, Cuzner ML, Hafler DA (1995) Expression of costimulatory molecules B7-1 (CD80), B7-2 (CD86), and interleukin 12 cytokine in multiple sclerosis lesions. J Exp Med 182:1985–1996

Cytokine and Anti-Inflammatory Drug Effects on Brain Trauma and Alzheimer's Disease Pathology in Transgenic Mice

P. H. Patterson

Summary

There is a continuous interaction between neurons, glia and inflammatory cells in the normal brain that is exacerbated following trauma or neurodegenerative disease. Cytokines mediate many of these interactions in injury and disease, and they may also play key roles in the developing and normal adult brain. Experiments using mice with targeted mutations in the gene for one of these cytokines, leukemia inhibitory factor (LIF), demonstrate that this protein is required for appropriate cell and behavioral responses to trauma in the nervous system. We find that LIF is induced by physical injury or chemically-induced inflammation, and in the absence of this cytokine, neuronal, astrocytic, microglial, neutrophil, macrophage and pain responses to these insults are strikingly altered. The specific effects of LIF deficiency depend critically on the site and type of injury, however. The inflammatory response can also be important in the establishment and/or progression of neurodegenerative disease. To study this we are using a transgenic mouse model of Alzheimer's disease (AD). Preliminary results indicate that stressing the animals can strongly up-regulate the microglial reaction around senile plaques.

Introduction

Leukemia inhibitory factor belongs to the neuropoietic cytokine family, which also includes ciliary neurotrophic factor (CNTF), oncostatin M (OSM), growth promoting activity, cardiotrophin-1, and interleukin-6 and -11 (IL-6 and IL-11) (Taga and Kishimoto 1997). All of these cytokines are pluripotent, having effects on proliferation and gene expression in a wide variety of cell types in culture. The challenge now is to sort out their various functions in vivo. Since all of these cytokines can utilize the same signal transducing receptor subunit, gp130, they can display identical biological effects on a number of different cell types. Thus, the issue of potential redundancy in function in vivo is a real one. Evidence for such redundancy (or cooperation) comes from comparison of mice in which single cytokines have been inactivated by gene targeting (knockout, KO) with receptor KO mice. While

Patterson/Kordon/Christen (Eds.)
Neuro-Immune Interactions in Neurologic and Psychiatric Disorders

IL-6, LIF and CNTF KO mice each live to maturity and reproduce, mice in which either the LIF binding receptor subunit or the gp130 transducing subunit has been inactivated die as embryos (Li et al. 1995; Ware et al. 1995). Additional evidence for cooperativity comes from study of motor neurons in LIF KO and CNTF KO mice. While motor function is minimally affected in the single gene KO mice, the double LIF/CNTF KO adult mice display a significant loss of motor neurons as they age, which is accompanied by a deficit in grip strength (Sendtner et al. 1996). These double KO mice also lose many more motor neurons following axotomy than do the single KO mice.

Despite this evidence for redundancy or cooperation, experiments with mice in which single neuropoietic cytokines have been inactivated have also demonstrated strong phenotypes in the adult nervous system. Inactivating LIF reveals that this cytokine is critical for a number of neuronal and glial functions in the adult mouse. For instance, despite a seemingly normal complement of neuronal phenotypes in the LIF KO brain, there are striking, sexually dimorphic alterations in glial phenotypes (Bugga et al. 1998). Moreover, LIF KO mice display a blunted stress response, with deficits in ACTH and corticosterone levels in response to immobilization or adjuvant injection (Chesnokova et al. 1998). In addition, in vivo experiments aimed at exploring potential therapeutic uses of LIF have shown that its application can rescue neurons from injury-induced death, enhance sprouting of processes, and regulate neuronal gene expression (reviewed by Murphy et al. 1997; Cheema et al. 1998; Kurek et al. 1998).

LIF Coordinates Neuroinflammatory Responses in the PNS and CNS

The use of LIF KO mice has also shown that LIF plays a central role in coordinating the neural and immune responses to injury and inflammation (Patterson 1994). One set of findings relates to neurogenic inflammation in peripheral tissues. Intraplantar injection of complete Freund's adjuvant (CFA) produces an inflammatory response characterized by local swelling, sensory hypersensitivity and a rapid up-regulation of IL-1β and nerve growth factor (NGF) in the skin, followed by up-regulation of the neuropeptides substance P and calcitonin gene-related peptide in sensory neurons. We found that LIF mRNA levels in the footpad increase following either CFA or IL-1 injection. Examination of LIF KO mice revealed that, while these mice display normal sensitivity to cutaneous mechanical and thermal stimulation, the degree of CFA-induced inflammation in the KO mice is strongly enhanced in spatial extent, amplitude, cellular infiltrate, IL-1β and NGF expression compared to wild-type (WT) or LIF heterozygote littermate mice (Banner et al. 1998). Conversely, local injection of low doses of recombinant LIF diminishes mechanical and thermal hypersensitivity induced by CFA, as well as the induction of IL-1β and NGF expression. Thus, the up-regulation of LIF induced by peripheral inflammation serves a key, early anti-inflammatory role. Moreover, exogenous LIF can be analgesic. These findings suggest that LIF merits testing in certain clinical settings involving neurogenic inflammation. It should be noted that Thompson et al. (1996), using young rats and higher doses for LIF injections, have drawn conclusions different from ours regarding the role of LIF in neuropathic pain. The role of the related cytokine, IL-6, has also been examined in inflammatory pain using IL-6 KO mice. Surprisingly, these mice exhibited a lower nociceptive threshold in the control condition but reduced hyperalgesia following carrageenan injection compared to WT mice (Xu et al. 1997).

Very different findings arose from study of the role of LIF in the neural and inflammatory responses to sciatic nerve crush or section. Here also, expression of LIF mRNA increases dramatically, primarily in Schwann cells (Banner and Patterson 1994; Curtis et al. 1994; Sun et al. 1994), and this rise is required for the appropriate neuronal response to trauma. The induction of several neuropeptides in sympathetic and sensory neurons in response to axotomy is not seen following such an injury to LIF KO nerves (Rao et al. 1993; Corness et al. 1996; Sun and Zigmond 1996a,b). We have also been investigating how LIF may be involved in the inflammatory response and find that LIF KO mice display a markedly deficient initial response to sciatic nerve crush (Patterson et al. 1997). Compared to WT or LIF heterozygote nerves, LIF KO nerves exhibit fewer neutrophils (identification based on morphology; 35% of WT; $p<.0001$), macrophages (based on morphology; 61% of WT; $p<.0001$) and mast cells (58% of WT as counted by morphological appearance; 55% of WT as counted as serotonin$^+$ cells; $p<.0001$) at the site of injury 24 hours following crush. At later time points, the difference between mutant and WT responses becomes progressively smaller, indicating that the response in the mutant is delayed rather than abolished.

We have examined whether this effect of LIF on inflammatory cell infiltration could be a direct, chemotactic one. Using a microchamber assay with mouse peritoneal macrophages, we find that LIF has significant chemotactic activity, displaying a classical bell-shaped dose-response curve, with peak activity at 1 nM LIF (Patterson et al. 1997). Interestingly, other members of this cytokine family (IL-6, IL-11, CNTF and OSM) did not display this activity when assayed in the same manner. To further examine the in vivo relevance of these results, we incubated injured sciatic nerves in culture media for 24–48 hours and found significant chemotactic activity for peritoneal macrophages. Moreover, media conditioned by injured nerves from LIF KO mice display significantly less chemotactic activity than media conditioned by WT nerves (G.K. Tofaris and P.H. Patterson, unpublished).

In sum, several lines of evidence support the idea that LIF is directly chemotactic for infiltrating inflammatory cells: 1) LIF has chemotactic activity for macrophages in a microchamber assay, 2) nerves from LIF KO mice display fewer inflammatory cells in vivo following injury than do WT nerves, and 3) injured nerves release a chemotactic activity in vitro and the level of this activity is reduced in LIF KO nerves.

The finding that the initial inflammatory response in injured nerve is reduced in the absence of LIF indicates a *pro*-inflammatory role for this cytokine. This conclusion is contradictory to that derived from the CFA studies in skin reviewed above. In the latter case, the inflammatory response was much greater in the absence of LIF, and application of exogenous LIF reduced inflammatory pain as well as IL-1β and NGF, which strongly argues for an *anti*-inflammatory role for LIF. Two of the differences that may underlie these opposing observations are, first, that two distinct types of trauma were used (physical injury vs. CFA), and second, the trauma was introduced in two very different locations (nerve vs. skin). Further experimentation should be able to determine which of these differences gives rise to the contradictory effects of LIF. For instance, physical injury can be studied in the skin. Our favored current hypothesis is that the skin contains a cell type that is not present in nerve, and that this cell reacts to the rise in LIF by producing a strong anti-inflammatory agent that overcomes the normally pro-inflammatory effect of LIF.

Perturbation studies on other tissues have also led to contrasting results. Injection of high doses of LIF in the mouse ear pinna increases skin thickness and stimulates leukocyte infiltration and IL-1α levels. Injection of low doses did not have this effect, however (MacKenzie et al. 1996). Thus, effects seen with KO mice, involving the removal of normal levels of cytokine, may differ significantly from effects seen at abnormally high levels following cytokine injection. Fortunately, in our study with CFA, the LIF injection results complemented the data from the LIF KO experiment. In a different preparation, we found that LIF rises in alveolar fluid in response to intratracheal LPS injection and that co-injection of LIF with LPS reduces TNF secretion and neutrophil invasion (Ulich et al. 1994). Exogenous LIF inhibits the development of experimental atherosclerosis in rabbit (Moran et al. 1994). In contrast, injection of LIF in monkeys elicits an acute phase reaction (Ryffel 1993). Passive immunization against LIF protects mice from lethal doses of endotoxin (Block et al. 1993). These and related studies on the neuropoietic cytokine family and inflammation are reviewed by Gadient and Patterson (1999).

Many of our findings with LIF in peripheral nerve also apply to CNS trauma. We have used surgical injury to the cerebral cortex of LIF KO mice

Table 1. Cytokine dependence of microglial and astrocyte reaction 24 hours following cortical injury[a]

	LIF KO	IL-6 KO
Mac-1:	49.9%	55.9
	$p<0.001$	0.002
F4/80:	59.7	75.2
	$p<0.003$	0.04

[a] Scanned optical density of immunohistochemical staining, expressed as % of WT. Not shown: Data for LIF heterozygote do not differ from WT, and data for double LIF/IL-6 KO does not differ significantly from those of the LIF KO.

to begin to investigate the role of this cytokine in the CNS. Expression of LIF mRNA is rapidly and widely induced in the normal brain following trauma, and this induction is highly localized to astrocytes (Banner et al. 1997). Expression of LIF mRNA also occurs in the spinal cord following surgical L.R. Banner and P.H. Patterson, unpublished) or excitotoxic lesion (Kurek et al. 1998). Moreover, we also find that the initial inflammatory cell infiltration into injured cortex is significantly slower in the absence of LIF (J. Han and P.H. Patterson, unpublished). For comparison, we studied the inflammatory reaction following surgical trauma in WT, LIF heterozygote and LIF KO as well as in IL-6 KO brains. We used optical image scanning of stained sections to quantitate the differences in Mac-1 and F4/80 staining of microglia and macrophages, the latter antibody highlighting a large subset of Mac-1^+ microglia. Some of these data are presented in Table 1. At 24 hours post-lesion, both immunohistochemical markers reveal that the microglial reaction around the lesion area in the LIF KO cortex is about half of that found in WT or heterozygote. This difference is highly significant ($p<0.001$). The microglial reaction in the IL-6 KO cortex is also less than the WT, but the difference is less dramatic, the KO reaction being 25–45% lower than WT ($p<0.04$ and 0.002 for the two markers). Perhaps surprisingly, in the one double LIF/IL-6 KO mouse examined thus far, the deficit in microglial reaction is no greater than in the LIF KO case (data not shown). Similar overall findings were obtained when the data were quantified as the number of cells stained with each of the three antibodies in each genotype (data not shown).

Our results with the IL-6 KO mouse are broadly similar to those of Klein et al. (1997), who studied the facial nucleus following axotomy of the motor neurons that reside there. They observed a striking deficit in GFAP induction and a much smaller decrease in $\alpha_M\beta_2^+$ microglia in the injured IL-6 KO compared to injured WT. In their study as well as ours, littermates are used as the controls in an attempt to minimize strain background differences when comparing genotypes. Several groups have taken the alternative approach of over-expressing IL-6 (as well as IL-3, tumor necrosis factor (TNF)α, transforming growth factor (TGF)β or interferon (IFN)α) in the brain using either neuron- or astrocyte-specific promoters (e.g., Campbell et al. 1998; Di Santo et al. 1996; Mattson et al. 1997). Not surprisingly, these powerful inflammatory agents evoke astrogliosis and microglial activation. Another neuropoietic cytokine family member, CNTF, is increased following brain trauma and this protein can elicit astrocyte activation when injected into the brain (Murphy et al. 1997; Hudgins and Levison 1998). Evaluation of CNS pathology in the absence of CNTF has not yet been reported, however.

Neuropoietic Cytokines and Seizure

Another form of trauma to the brain is the dramatic increase in neuronal activity caused by seizure. In both human patients and a variety of animal models, seizure can induce a number of neuronal and glial genes as well as astrocyte activation, axonal sprouting, and neuronal cell death. To investigate the role of cytokines in the pathological as well as recovery phases of epilepsy, we have characterized the expression of LIF, OSM, CNTF and CT-1 in the rat hippocampus following a pilocarpine-induced seizure. In this model, a single, prolonged seizure is induced by i.p. injection of pilocarpine. Using a RNAse protection assay, we find that the levels of mRNAs for all four neuropoietic cytokines are increased by seizure, but the changes in CT-1 and CNTF are small (Jankowsky and Patterson 1998). In contrast, LIF mRNA, normalized to the housekeeping enzyme glyceraldehyde phosphate dehydrogenase, increases 30-fold by 12 hours following pilocarpine injection, and OSM is increased 10-fold by 2 hours. Unlike LIF, which returns to baseline by three days following seizure and stays low, OSM mRNA levels fall towards baseline and then display a reproducible second peak of induction at seven days post-seizure. These two distinct peaks of expression suggest that OSM may have more than one function in seizure onset and recovery. In addition, Minami et al. (1991) found that LIF and IL-6 mRNAs are increased by kainic acid injection at a convulsive dose.

Using in situ hybridization and immunolabeling for neuronal and glial markers, we find that the cell types expressing LIF and OSM mRNAs at their early peaks are different. As in the case of surgical trauma, seizure induces LIF up-regulation in astrocytes. In addition, cells in cortical blood vessels and cells associated with the meninges are frequently LIF^+. Thus, strong activation of neurons during seizure induces LIF expression not only in astrocytes but also in cells outside and/or within the blood brain barrier. The identity of the latter cells is being pursued. In contrast to LIF, the expression of OSM mRNA at two hours following pilocarpine injection is induced not in astrocytes but rather in cells with the features of interneurons (location, size and occasional staining with neurofilament antibodies, $GFAP^-$). To our knowledge, this is the first finding of OSM expression in the nervous system, and it suggests that OSM could play a role in the signaling events that attempt to reduce the immediate effects or to propagate the seizure. We have not yet been able to determine the cellular localization of the OSM mRNA that is induced again at seven days following pilocarpine injection, but this timing suggests a function in the recovery process that follows seizure.

We are now using LIF KO mice to examine the function of LIF in post-seizure events. Since LIF is induced in astrocytes shortly after seizure, and it is clear that LIF can regulate GFAP immunolabeling in astrocytes in vivo (Bugga et al. 1998), this is a logical target gene for LIF action in this model. Various neuropeptides and other neuronal genes are also up-regulated by seizure (e.g., Lynch et al. 1996) as well as by LIF in other contexts, including trauma, so these are also good candidate genes. Given that electroconvulsive shock therapy can be an effective therapeutic intervention in severe depression, further examination of the molecular signals that regulate seizure and its effects are warranted.

Inflammation, Cytokines and Alzheimer's Disease

Several lines of evidence point to inflammation as an important feature of Alzheimer's disease (AD) (McGeer and McGeer 1995; Patterson 1995). The senile plaques that are characteristic of this disease are sites of microglial activation, and expression of a number of cytokines including TNFα, macrophage colony stimulating factor, IL-1, IL-3 and IL-6, as well as acute phase and complement proteins is also present at these sites (Griffin et al. 1995; Huell et al. 1995; Sheng et al. 1996; Mattson et al. 1997; Yan et al. 1997; Griffin et al. 1998). Such inflammatory markers are prominent in developing plaques, arguing that this is not an end-stage reaction (Sasaki et al. 1997; Sheng et al. 1997). In addition, these markers are also induced by head trauma, an injury that increases the risk of AD (Griffin et al. 1995b). Culture experiments have shown that there are a number of interactions between β-amyloid and various components of the inflammatory reaction (Rogers et al. 1992; Pachter 1997). Another cytokine of relevance is TGFβ1, which is induced by brain injury and is elevated in AD brains (Mattson et al. 1997). Moreover, transgenic mice engineered to over-express both TGFβ1 and amyloid-precursor protein (APP) display accelerated deposition of β-amyloid compared to littermates over-expressing just APP (Wyss-Coray et al. 1997). Given that many cytokines can display both trophic and pathological effects at various concentrations, however, results from transgenic mice over-expressing a cytokine do not necessarily reveal the normal function of that cytokine. Intraventricular infusion of the APP peptide 1–40 results in amyloid deposits that are greatly potentiated by simultaneous injection in the brain parenchyma of TGFβ (Frautschy et al. 1996). In fact, evoking inflammation in the brain non-specifically by injection of lipopolysaccharide either in the ventricles or peripherally can alter APP expression in the brain (Brugg et al. 1995; Hauss-Wegrzyniak et al. 1998), and many interactions have been reported between APP and cytokines in culture experiments (McGeer and McGeer 1997).

The therapeutic implications of the association between AD and localized inflammation are further supported by a number of epidemiological studies showing that sustained use of non-steroidal anti-inflammatory drugs (NSAIDs) is associated with delayed onset and reduced risk of AD (e.g., Breitner, 1996; McGeer et al. 1996). There is also a recent report of a possible genetic link to the HLA immune response locus; the A2 allele, or a closely linked gene, is elevated in frequency in early- and late-onset AD (Payami et al. 1997). In addition, results from several small clinical trials suggest that anti-inflammatory drugs can have efficacy in the treatment of memory loss in AD patients (Fudenberg and Singh 1988; McGeer and Rogers 1992; Leszek and Gasiorowski 1996).

To explore this possibility in more detail, we are using one of the mouse models of AD (Duff 1998). This transgenic mouse over-expresses the isoform of human APP containing the Swedish mutation that causes AD. This results in mice with an age-dependent increase in APP deposition that is accompanied by plaques and behavioral deficits (Hsiao et al. 1996). To ask whether

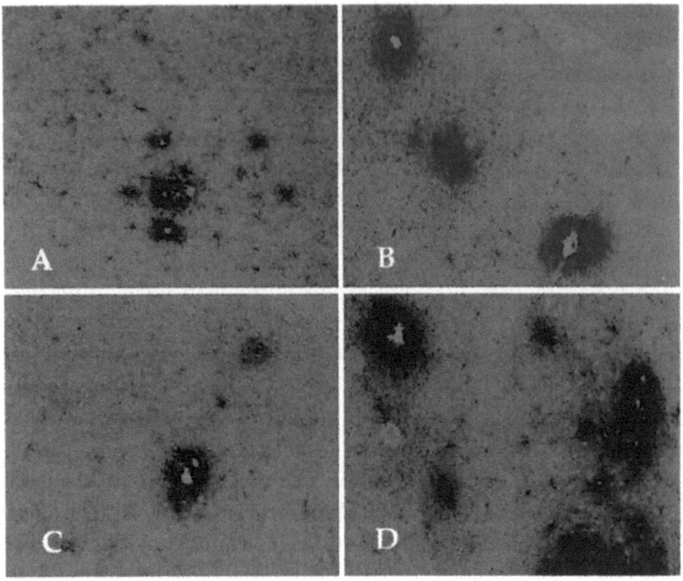

Fig. 1. Effects of saline and anti-inflammatory drug injection on the microglial reaction surrounding plaques in a mouse AD model. Transgenic mice bearing the mutant human gene that can cause AD were injected i.p. twice a week, beginning at age 9 months, and killed at 13 months. The panels depict typical plaques from uninjected mice (A), or mice injected with saline (B), the NSAID meloxicam (C) or the corticosteroid dexamethasone (D) (the latter two given at 10 μg and 1.5 μg per injection, respectively). Brain sections were stained with thioflavin to visualize plaques and with Mac-1 to label microglia. There is a strong increase in microglial immunostaining caused by injection of any of the substances, suggesting that stress activates plaque-associated inflammation. There is also no clear effect of the anti-inflammatory drugs on this inflammation

anti-inflammatory drugs can affect plaque formation and microglial activation in this model, we are injecting dexamethasone, a broadly acting corticosteroid, and meloxicam, an NSAID that preferentially inhibits COX-2, a key prostaglandin-producing enzyme induced by inflammation (Furst 1997; Vane and Botting 1997). Corticosteroids are generally more potent but less specific anti-inflammatory drugs than NSAIDs. There is some evidence that COX-2 is up-regulated in AD (Lukiw and Bazan 1997; Oka and Takashima 1997; Aisen and Pasinetti 1998). In the first series of experiments, the APP transgenic mice were injected i.p., twice a week, beginning at age nine months, and they were killed at 13 months.

In this work (R. Lahav, R. Zaman, S. Tolaney and P.H. Patterson, unpublished), the first finding of interest is that injection of saline, as a control for drug injection, itself causes a marked increase in microglial reaction around the plaques (Figure 1). We hypothesize that this striking effect is induced by the stress of injection twice per week. This result has potential clinical implications for AD. The second finding is that this drug treatment protocol did not further modify the microglial reaction caused by the plaques and the stress of injection (Figure 1). The third finding is that this drug treatment protocol did not result in a major reduction in the density of plaques. There is, however, a slight effect of meloxicam. In the first set of mice, the mean APP^+ plaque density ± S.E.M. was 2.95 ± 0.22 for control mice and 2.38 ± 0.16 and 2.66 ± 0.21 for meloxicam- and dexamethasone-treated mice, respectively. Two-tailed T tests reveal a significance value of $p < 0.044$ for this 19% difference between control and meloxicam-treated mice. Current experiments are aimed at determining if altering the treatment protocol can affect the degree of this difference.

The only relevant human studies of which we are aware recently reported mixed neuropathological findings on the effects of anti-inflammatory drugs. Mackenzie and Munoz (1998) found that the prevalence of senile plaques is not different between non-AD elderly patients treated with NSAIDs and controls. However, the NSAID-treated patients displayed three-fold fewer activated microglia than controls. The latter result is different from ours. In contrast, in a preliminary communication Halliday et al. (1997) report that glucocorticoid or NSAID use by AD patients correlates with significantly more microglia and more diffuse and cored plaques than non-drug using AD or control subjects. These mixed results from small human studies provide further impetus to examine these drugs in the transgenic mouse model where drug treatment can be closely controlled.

In animal studies, Castano et al. (1996) observed that dexamethasone treatment increased the number of $F4/80^+$ microglia in the crushed rat optic nerve, while Giulian et al. (1989) reported that dexamethasone had no effect on inflammatory cell infiltration following traumatic brain injury. On the other hand, in the axotomized facial nucleus, this drug was found to have no effect on markers of microglial activation or on their morphology but did inhibit the induction of MHC class II expression (Kiefer and Kreutzberg 1991). Dexamethasone was also found to inhibit the enhancement of OX-42 staining of microglia caused by K^+-induced spreading depression (Caggiano and Kraig 1996). In addition, Netland et al. (1998) recently reported that the microglial response to chronic injection of βA was attenuated by injection of indomethacin. It may also be of interest that AD patients display a deficit in comparison to normal elderly controls in the corticosteroid regulation of hippocampal glucose (de Leon et al. 1997).

To further examine this issue, as well as the question of how to reduce plaque formation, we will test additional anti-inflammatory drugs and new treatment protocols, including a different method of drug delivery. We have also instituted a breeding program that is producing APP^+ transgenic mice that lack particular neuropoietic cytokines while preserving the essential features of the APP mouse genetic background. This experiment will test whether these cytokines are essential for plaque development and for the inflammatory events that accompany and promote them.

Acknowledgments. Lisa Banner, Jing Han, Joanna Jankowsky, Ronit Lahav, Shigeki Sugiura, George Tofaris, Sara Tolaney and Rumina Zaman carried out the current work cited from our group. These projects are being supported by the NINDS, the Della Martin and the McGrath Foundations, Caltech Summer Undergraduate Research Fellowships, and the Human Frontiers in Science Program. I also acknowledge the technical and clerical support of Doreen McDowell and Floreen Rooks, and thank Jian Zhong, Sigrid Schwarz, Ronit Lahav and Joanna Jankowsky for their helpful comments on the manuscript.

References

Aisen PS, Pasinetti GM (1998) Glucocorticoids in Alzheimer's disease. Drugs Aging 12:1–6

Banner LR, Patterson PH (1994) Major changes in the expression of the mRNAs for CDF/LIF and its receptor after injury to adult peripheral nerves and ganglia. Proc Natl Acad Sci, USA 91:7109–7113

Banner LR, Moayeri NN, Patterson PH (1997) Leukemia inhibitory factor is expressed in astrocytes following cortical brain injury. Exp Neurol 147:1-9

Banner LR, Patterson PH, Allchorne A, Poole S, Woolf CJ (1998) Leukemia inhibitory factor is an anti-inflammatory and analgesic cytokine. J Neurosci 18:5456–5462

Block MI, Berg M, McNamara MJ, Norton JA, Fraker DL, Alexander HR (1993) Passive immunization of mice against D factor blocks lethality and cytokine release during endotoxemia. J Exp Med 178:1085–1090

Breitner JCS (1996) The role of anti-inflammatory drugs in the prevention and treatment of Alzheimer's disease. Annu Rev Med 47:401–411

Brugg B, Dubreuil LY, Huber G, Wollman EE, Delhaye-Bouchaud N, Mariani J (1995) Inflammatory processes induce β-amyloid precursor protein changes in mouse brain. Proc Natl Acad Sci, USA 92:3032–3035

Bugga L, Gadient RA, Kwan K, Stewart CL, Patterson PH (1998) Analysis of neuronal and glial phenotypes in brains of mice deficient in leukemia inhibitory factor. J Neurobiol 36:509–524

Caggiano AO, Kraig RP (1996) Eicosanoids and nitric oxide influence induction of reactive gliosis from spreading depression in microglia but not astrocytes. J Comp Neurol 369:93–108

Campbell IL, Stalder, AK, Akwa Y, Pagenstecher, A and Asenio, VC (1998) Transgenic models to study the actions of cytokines in the central nervous system. Neuroimmunomodulation 5:126–135

Castano A, Lawson LJ, Fearn S, Perry VH (1996) Activation and proliferation of murine microglia are insensitive to glucocorticoids in Wallerian degeneration. Eur J Neurosci 8:581–588

Cheema SS, Arumugam D, Murray SS, Bartlett PF (1998) Leukemia inhibitory factor maintains choline acetyltransferase expression *in vivo.* NeuroReport 9:363–366

Chesnokova V, Auernhammer CJ, Shlomo M (1998) Murine leukemia inhibitory factor gene disruption attenuates the hypothalamo-pituitary-adrenal axis stress response. Endocrinology 139:2209–2216

Corness J, Shi T-J, Xu Z-Q, Brulet P, Hokfelt T (1996) Influence of leukemia inhibitory factor on galanin/GMAP and neuropeptide Y expression in mouse primary sensory neurons after axotomy. Exp Brain Res 112:79–88

Curtis R, Scherer S, Somogyi R, Adryan KM, Ip N, Zhu Y, Lindsay RM, DiStefano PS (1994) Retrograde axonal transport of LIF is increased by peripheral nerve injury: correlation with increased LIF expression in distal nerve. Neuron 12:191–204

de Leòn MJ, McRae T, Rusinek H, Convit A, De Santi S, Tarshish C, Golomb J, Volkow N, Daisley K, Orentreich N, McEwen B (1997) Cortisol reduces hippocampal glucose metabolism in normal elderly, but not in Alzheimer's disease. J Clin Endocrinol Metab 82:3251–3259

Di Santo E, Alonzi T, Fattori E, Poli V, Ciliberto G, Sironi M, Gnocchi P, Ricciardi-Castagnoli P, Ghezzi P (1996) Overexpression of interleukin-6 in the central nervous system of transgenic mice increases central but not systemic proinflammatory cytokine production. Brain Res 740:239–244

Duff K (1998) Transgenic models for Alzheimer's disease. Neuropathol Appl Neurobiol 24:81–103

Frautschy SA, Yang F, Calderon L, Cole GM (1996) Rodent models of Alzheimer's disease: rat Aβ infusion approaches to amyloid deposits. Neurobiol Aging 17:311–321

Fudenberg HH, Singh V (1988) Alzheimer's 'syndrome': prognosis of subsets with different etiology and preliminary effects of immunotherapy. Drug Dev Res 15:165–174

Furst DE (1997) Meloxicam: selective COX-2 inhibition in clinical practice. Sem Arthritis Rheum 26:21–27

Gadient, RA and Patterson, PH (1999) Leukemia inhibitory factor, interleukin-6 and other neuropoietic cytokines in inflammation and injury. Stem Cells 17:127–137

Giulian D, Chen J, Ingeman JE, George JK, Noponen M (1989) The role of mononuclear phagocytes in wound healing after traumatic injury to adult mammalian brain. J Neurosci 9:4416–4429

Griffin WST, Sheng JG, Roberts GW, Mrak RE (1995a) Interleukin-1 expression in different plaque types in Alzheimer's disease: significance in plaque evolution. J Neuropathol Exp Neurol 54:276–281

Griffin WST, Sheng JG, Gentleman SM, Graham DI, Mrak RE, Roberts GW (1995b) Microglial interleukin-1α expression in human head injury: correlations with neuronal and neuritic β-amyloid precursor protein expression. Neurosci Lett 176: 133–136

Griffin WST, Sheng JG, Mrak RE (1997) Inflammatory pathways: implications in Alzheimer's disease. In: Wasco W, Tanzi RE (eds) Molecular Mechanisms of Dementia. Humana Press, Totowa, New Jersey, pp. 169–176

Griffin WST, Sheng JG, Royston MC, Gentleman SM, McKenzie JE, Graham DI, Roberts, GW, Mrak RE (1998) Glial-neuronal interactions in Alzheimer's disease: the potential role of a 'cytokine cycle' in disease progression. Brain Pathol 8:65–72

Halliday GM, McCann H, Brooks WS, Brea GA (1997) Inflammation in Alzheimer's disease: anti-inflammatory medications increase postmortem neuropathology. Brain Pathol 7:1054

Hauss-Wegrzyniak B, Dobrzanski P, Stoehr JD, Wenk GL (1998) Chronic neuroinflammation in rats reproduces components of the neurobiology of Alzheimer's disease. Brain Res 780:294–303

Hsiao K, Chapman P, Nilsen S, Eckman C, Harigaya Y, Younkin S, Yang F, Cole G (1996) Correlative memory deficits, Aβ elevation, and amyloid plaques in transgenic mice. Science 274:99–102

Hudgins SN, Levison SW (1998) Ciliary neurotrophic factor stimulates astroglial hypertrophy *in vivo* and *in vitro*. Exp Neurol 150:171–182

Huell M, Strauss S, Volk B, Berger M, Bauer J (1995) Interleukin-6 is present in early stages of plaque formation and is restricted to the brains of Alzheimer's disease patients. Acta Neuropathol 89:544–551

Jankowsky JL, Patterson PH (1999) Differential regulation of cytokine expression following pilocarpine-induced seizure. Exper Neurol, in press

Kiefer R, Kreutzberg W (1991) Effects of dexamethasone on microglial activation in vivo: selective downregulation of major histocompatibility complex class II expression in regenerating facial nucleus. J Neuroimmunol 34:99–108

Klein MA, Moller JC, Jones LL, ,Bluethmann H, Kreutzberg GW, Raivich G (1997) Impaired neuroglial activation in interleukin-6 deficient mice. Glia 19:227–233

Kurek JB, Bennett TM, Bower JJ, Muldoon CM, Austin L (1998) Leukaemia inhibitory factor (LIF) production in a mouse model of spinal trauma. Neurosci Lett 249:1–4

Leszek J, Gasiorowski K (1996) Therapeutic efficacy of cyclophosphamide in Alzheimer's disease. Alzheimer's Res 2:43–46

Li M, Sendtner M, Smith A (1995) Essential function of LIF receptor in motor neurons. Nature 378:724–727

Lukiw WJ, Bazan NG (1997) Cyclooxygenase 2 RNA message abundance, stability, and hypervariability in sporadic Alzheimer neocortex. J Neurosci Res 50:937–945

Lynch MW, Rutecki PA, Sutula TP (1996) The effects of seizures on the brain. Curr Opin Neurol 9:97–102

MacKenzie IRA, Munoz DG (1998) Nonsteroidal anti-inflammatory drug use and Alzheimer-type pathology in aging. Neurology 50:986–990

Mattson MP, Barger SW, Furukawa K, Bruce AJ, Wyss-Coray T, Mark RJ, Mucke L (1997) Cellular signaling roles of TGFβ, TNFα and βAPP in brain injury responses and Alzheimer's disease. Brain Res Rev 23:47–61

McGeer P, Rogers J (1992) Anti-inflammatory agents as a therapeutic approach to Alzheimer's disease. Neurology 42:447–449

McGeer PL, McGeer EG (1995) The inflammatory response system of brain: implications for therapy of Alzheimer and other neurodegenerative diseases. Brain Res 21:195–218

McGeer PL, Schulzer M, McGeer EG (1996) Arthritis and anti-inflammatory agents as possible protective factors for Alzheimer's disease: a review of 17 epidemiologic studies. Neurology 47:425–432

McGeer EG, McGeer PL (1997) Inflammatory cytokines in the CNS. Possible role in the pathogenesis of neurodegenerative disorders and therapeutic implications. Pharmacol Pathophysiol 7:214–228

McKenzie RC, Pagilia D, Kondo S, Sauder DN (1996) A novel endogenous mediator of cutaneous inflammation: leukemia inhibitory factor. Acta Derm Venereol (Stockh) 76:111–114

Minami M, Kuraishi Y, Satoh M (1991) Effects of kainic acid on messenger RNA levels of IL-1β, IL-6, TNFα and LIF in the rat brain. Biochem Biophys Res Commun 176:593–598

Moran CS, Campbell JH, Simmons DL, Campbell GR (1994) Human leukemia inhibitory factor inhibits development of experimental atherosclerosis. Arterioscler Throm 14:1356–1363

Murphy M, Dutton R, Koblar S, Cheema S, Bartlett P (1997) Cytokines which signal through the LIF receptor and their actions in the nervous system. Progr Neurobiol 52:355–378

Netland EE, Newton JL, Majocha RE, Tate BA (1998) Indomethacin reverses the microglial response to amyloid β-protein. Neurobiol Aging 19:201–204

Oka A, Takashima S (1997) Induction of cyclo-oxygenase 2 in brains of patients with Down's syndrome and dementia of Alzheimer type: specific localization in affected neurones and axons. NeuroReport 8:1161–1164

Pachter JS (1997) Inflammatory mechanisms in Alzheimer disease: the role of beta-amyloid/glial interactions. Mol Psychiat 2:91–95

Patterson PH (1994) Leukemia inhibitory factor, a cytokine at the interface between neurobiology and immunology. Proc Natl Acad Sci USA 91:7833–7835

Patterson PH (1995) Cytokines in Alzheimer's disease and multiple sclerosis. Curr Opin Neurobiol 5:642-646

Patterson PH, Kou SY, Sugiura S, Lahav R, Banner LR (1997) LIF coordinates neuronal and inflammatory responses to nerve injury. Soc Neurosci Abstr 23:393.33

Payami H, Schellenberg GG, Zareparsi S, Kaye J, Sexton GJ, Head MA, Matsuyama SS, Jarvik LF, Miller B, McManus DQ, Bird TD, Katzman R, Heston L, Norman D, Small GW (1997) Evidence for association of HLA-A2 allele with onset age of Alzheimer's disease. Neurology 49:512–518

Rao MS, Sun Y, Escary JL, Perreau J, Tresser S, Patterson PH, Zigmond RE, Brulet P, Landis SC (1993) Leukemia inhibitory factor mediates an injury response but not a target-directed developmental transmitter switch in sympathetic neurons. Neuron 11:1175–1185

Rogers J, Cooper NR, Webster S, Schultz J, McGeer PL, Styren SD, Civin WH, Brachova L, Bradt B, Ward P (1992) Complement activation by beta-amyloid in Alzheimer disease. Proc Natl Acad Sci USA 89:10016–10020

Ryffel B (1993) Pathology induced by leukemia inhibitory factor. Int Rev Exp Pathol 34B:69–72

Sasaki A, Yamaguchi H, Ogawa A, Sugihara S, Nakazato Y (1997) Microglial activation in early stages of amyloid β protein deposition. Acta Neuropathol 94:316–322

Sendtner M, Gotz R, Holtmann B, Escary J-L, Masu Y, Carroll P, Wolf E, Brem G, Brulet P, Thoenen H (1996) Cryptic physiological trophic support of motoneurons by LIF revealed by double gene targeting of CNTF and LIF. Curr Biol 6:686–694

Sheng JG, Ito K, Skinner RD, Mrak RE, Rovnaghi CR, Van Eldik LJ, Griffin WST (1996) In vivo and in vitro evidence supporting a role for the inflammatory cytokine interleukin-1 as a driving force in Alzheimer pathogenesis. Neurobiol Aging 17:761–766

Sheng JG, Mrak RE, Griffin ST (1997) Neuritic plaque evolution in Alzheimer's disease is accompanied by transition of activated microglia from primed to enlarged to phagocytic forms. Acta Neuropathol 94:1–5

Sun Y, Rao M, Zigmond RE, Landis SC (1994) Regulation of vasoactive intestinal peptide expression in sympathetic neurons in culture and after axotomy: the role of cholinergic differentiation factor/leukemia inhibitory factor. J Neurobiol 25:415–430

Sun Y, Zigmond RE (1996a) Leukaemia inhibitory factor induced in the sciatic nerve after axotomy is involved in the induction of galanin in sensory neurons. Eur J Neurosci 8:2213–2220

Sun Y, Zigmond RE (1996b) Involvement of leukemia inhibitory factor in the increases in galanin and vasoactive intestinal peptide mRNA and the decreases in neuropeptide Y and tyrosine hydroxylase mRNA in sympathetic neurons after axotomy. J Neurochem 67:1751–1760

Taga T, Kishimoto T (1997) gp130 and the interleukin-6 family of cytokines. Annu Rev Immunol 15:797–819

Thompson SWN, Dray A, Urban L (1996) Leukemia inhibitory factor induces mechanical allodynia but not thermal hyperalgesia in the juvenile rat. Neuroscience 71:1091–1094

Ulich TR, Fann M-J, Patterson PH, Williams JH, Samal B, Del Castillo J, Yin S, Guo K, Remick DG (1994) Intratracheal injection of LPS and cytokines. V. LPS induces expression of LIF and LIF inhibits acute inflammation. Am Physiol Soc 267:442–446

Vane JR, Botting RM (1997) Mechanism of action of aspirin-like drugs. Sem Arthritis Rheum 26:2–10

Ware CB, Horowitz MC, Renshaw BR, Hunt JS, Liggitt D, Koblar SA, Fliniak BC, McKenna HJ, Papayannopoulou T, Thoma B, Cheng L, Donovan P, Peshon JJ, Bartlett PF, Willis CR, Wright BD, Carpenter MK, Davidson BL, Gearing DP (1995) Targeted disruption of the low-affinity lekemia inhibitory factor receptor gene causes placental, skeletal, neural and metabolic defects and results in perinatal death. Development 121:1283–1299

Wyss-Coray T, Masliah E, Mallory M, McConiogue L, Johnson-Wood K, Lin C, Mucke L (1997) Amyloidgenic role of cytokine TGF-β1 in transgenic mice and in Alzheimer's disease. Nature 389:603–606

Xu X-J, Hao J-X, Andell-Johnson S, Poli V, Bartfai T, Wiesenfeld-Hallin Z (1997) Nociceptive responses in interleukin-6-deficient mice to peripheral inflammation and peripheral nerve section. Cytokine 9:1028–1033

Yan SD, Zhu H Fu, Jin, Yan SF, Roher A, Tourtellotte WW, Rajavashisth T, Chen X, Godman GC, Stern D, Schmidt AM (1997) Amyloid-β peptide-receptor for advanced glycation endproduct interaction elicits neuronal expression of macrophage-colony stimulating factor: a proinflammatory pathway in Alzheimer disease. Proc Natl Acad Sci USA 94:5296–5301

In vivo Involvement of the Cytokine LIF During Lesion-Induced Renewal of Olfactory Sensory Neurons in Adult Mouse

S. Bauer, C. Mauduit, F. Jourdan, M. Benahmed, and E. Moyse

Introduction

Mammalian olfactory epithelium lines the deepest part of the nasal cavity at the level of olfactory turbinates and is made up predominantly of odorant-receptive neurons, each one projecting its unique axon to the ipsilateral olfactory bulb (Crews and Hunter 1994). These neurons have a limited lifspan, die spontaneously by apoptosis (Mahalik 1996), and are renewed throughout adult life by proliferation of local neuronal progenitors called globose basal cells (Fig. 1A; Monti-Graziadei and Graziadei 1979; Caggiano et al. 1994).

Recent *in vitro* studies of olfactory neurogenesis have identified a variety of growth factors and cytokines controlling proliferation or differentiation of cultured olfactory cells (Mahanthappa and Schwarting 1993; DeHamer et al. 1994; Farbman and Bucholtz 1996; MacDonald et al. 1996; Vawter et al. 1996; Goldstein et al. 1997; Satoh and Yoshida 1997a, b). *In situ* investigations have specified that some of the growth factors showing mitogenic activity *in vitro* are indeed present in olfactory epithelium of adult mammals (Deckner et al. 1993; Krishna et al. 1996). But there is no evidence regarding which mitogenic signals are effectively involved *in vivo* to induce the proliferation of olfactory neuronal progenitors.

We addressed this issue by using an experimental paradigm that passively enhances neuronal turnover in olfactory epithelium *in vivo*, i.e. ablation of the synaptic target, the olfactory bulb, or bulbectomy. This lesion sequentially triggers generalized apoptosis of sensory neurons (Michel et al. 1994) and then mitotic stimulation of globose basal cells (Schwartz-Levey et al. 1991), peaking, respectively, 36 hours and five days after injury (Fig. 1B). Since bulbectomy-induced mitotic stimulation is massive and synchronized at the level of the olfactory organ, physiological mitogenic signals should be detectable at the messenger RNA (mRNA) level by reverse transcription polymerase chain reaction (RT-PCR) in tissular extracts. We used a semi-quantitative RT-PCR approach that focused on the growth factors and cytokines that had been shown to stimulate *in vitro* proliferation of olfactory neuroblast cultures.

Patterson/Kordon/Christen (Eds.)
Neuro-Immune Interactions in Neurologic and Psychiatric Disorders

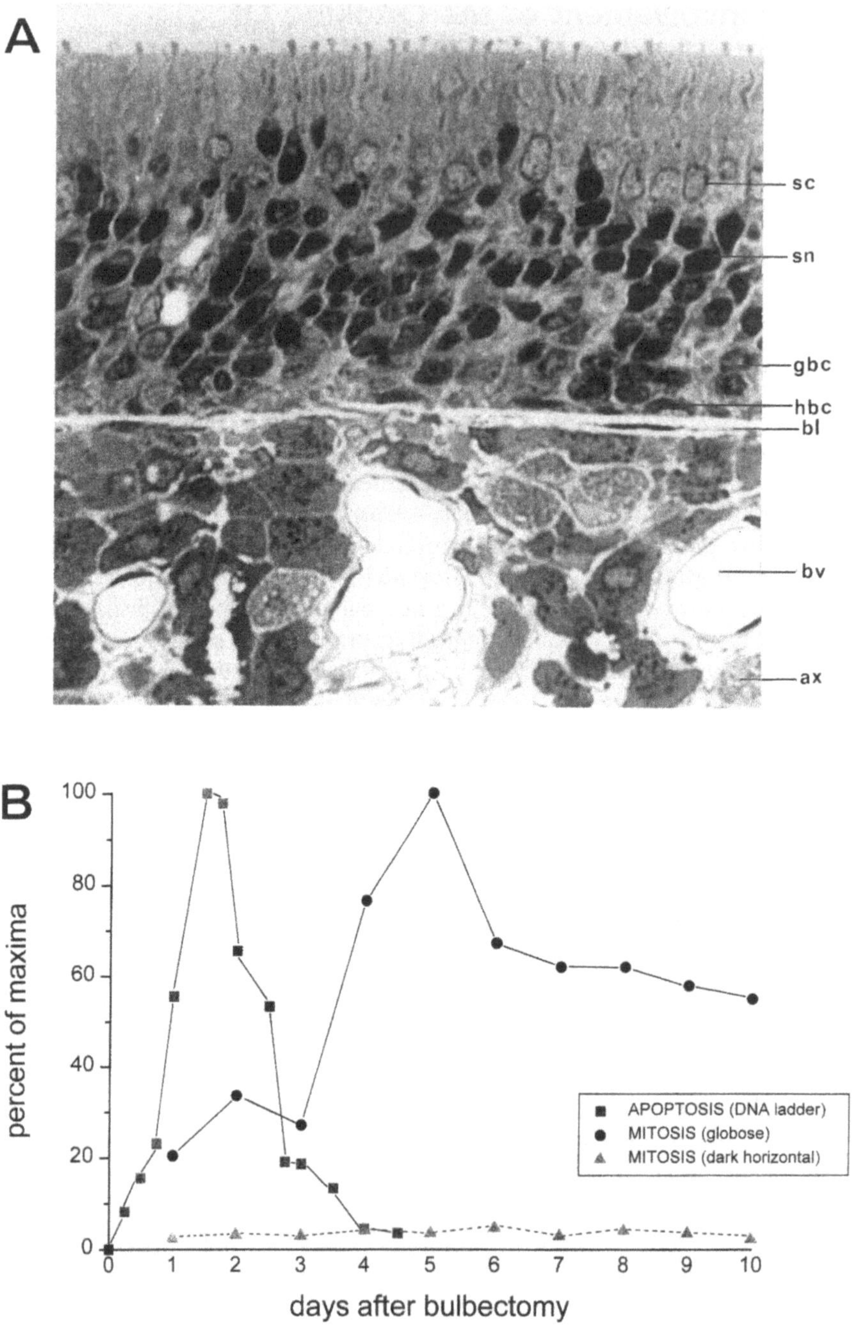
A
sc
sn
gbc
hbc
bl
bv
ax
B
100
80
60
40
20
0
percent of maxima
0
1
2
3
4
5
6
7
8
9
10
days after bulbectomy
APOPTOSIS (DNA ladder)
MITOSIS (globose)
MITOSIS (dark horizontal)

Fig. 13 A, B

Experimental Procedures

Adult, three-month-old mice of the C57 Bl/6J strain were obtained from IFFA-CREDO (France). In this study, we used 50 animals that were subjected to a bilateral bulbectomy, as described previously (Michel et al. 1994). Then animals were sacrificed by decapitation at different times ranging from eight hours to 10 days after surgery, for assessment of gene expression kinetics in the olfactory mucosa. The bulk of olfactory turbinates was immediately dissected out from mouse heads and stored at –80 °C.

Total cellular RNA was extracted from frozen tissues using TRIzol™ (guanidium isothyocianate-acidic phenol) from Life Technologies, following manufacturer's instructions. Briefly, each sample of frozen tissue was dry-crushed and received 1 ml of TRIzol™. Tubes were then mixed well at room temperature, and 0.5 ml of chloroform was added to differentially solubilize RNA. After recuperation of the acqueous phase, RNAs were precipitated with isopropanol (Sigma), washed in ethanol 70%, and dried at room temperature. Resultingt pellets were resuspended in RNAse-free water, and RNA concentration for each sample was determined spectrophotometrically by recording absorbance at 260 nm.

Reverse Transcription (RT) was conducted with 3 µg of total RNA from each sample in a 20 µl final reaction volume containing 200 U of Moloney Murine Leukemia Virus Reverse Transcriptase from Gibco in the buffer supplied with it (1X final), 5 µM random hexamere primers (Sigma), 0.25 mM of each nucleotide (dNTPs from Gibco), and dithiothreitol 0.01M. The mixture was heated for 60 min at 37 °C, then for five minutes at 95 °C for enzyme inactivation, and was finally diluted vol/vol with sterile water and stored at –20 °C.

PCRs were carried out with 1 µl (1/40) of total cDNA from each sample in a final reaction volume of 20 µl containing 0.25 U of Taq Polymerase (Promega) in manufacturer's buffer (1X final) and MgC12 (1.5mM) (Amersham) l µM of primers, 0.1 mM of dNTPs, and 0.0375 µCi of [^{33}P]-dATP. Forward and reverse primers were defined in separate exons to avoid any bias due to residual genomic contamination (Grey et al. 1983; Chatelain et al. 1995; Michel et al. 1997). We first determined optimal PCR conditions for each primer pair, the annealing temperature and the minimum number of

Fig. 1. **(A)** Light microscopic structure of olfactory epithelium in control adult mouse, as observed on a semi-thin (1-µm thick) section of prefixed olfactory organ. Olfactory epithelium lines up nasal cavity (at top) and is separated by a basal lamina (bl) from underlying lamina propria that harbors axon bundles of olfactory nerve (ax), blood vessels (bv) and Bowman's glands. The pseudostratified aspect of olfactory epithelium is due to laminar distribution of the respective nuclei of horizontal (hbc) and globose (gbc) basal cells, olfactory sensory neurons (sn) and supporting cells (sc). Bar, 25 µm **(B)** Kinetics of bulbectomy-induced apoptosis and proliferation in mouse olfactory epithelium *in vivo*, assessed, respectively, by DNA internucleosomal fragmentation (Michel et al. 1994) and autoradiography of *in vivo* incorporated ^{3}H-thymidine (Schwartz-Levey et al. 1991). Graphs were redrawn after normalization from cited bibliographic sources

PCR cycles necessary to detect radiolabeled reaction products at a unique, expected, molecular size. All controls were systematically done to check for contaminations. By using a MJ-Research thermocycler, the PCR mixture was first heated at 94 °C for 5 min, then 17–25 cycles of 92 °C for 30 sec, annealing temperature for 30 sec, 72 °C for 30 sec, and finally 72 °C for five min. Amplification products were then stored at 4 °C and separated on a poly-acrylamide gel electrophoresis system harboring 50 samples at once (Protean-2, Bio-Rad). We used 8% poly-acrylamide gels that were vacuum dried, and then processed for autoradiography with Bio-max films (Kodak). PCR-amplified products were checked by restriction enzymes. Moreover, for all primers, no amplification was observed when PCR was performed on RNA preparations.

The use of radioactive nucleotide allows relative quantification of PCR amplification products among all samples processed in the same assay. Intensities of autoradiographic bands were estimated by densitometric scanning using the Bio-Image Intelligent Quantifier program (Millipore). The data were expressed as target gene over β-actin mRNA ratio. PCR analysis were carried out from the logarithmic phase of amplification.

Results

Leukemia inhibitory factor (LIF) induction began as early as eight hours after ablation of olfactory bulbs and reached its maximum 24 hours after surgery, i.e., before neuronal apoptosis peak in the olfactory epithelium. LIF mRNA returned to control level by three to five days (Fig. 2). LIF-R mRNA rose up concomitantly during the first day after bulbectomy, and fell below control level until the end of the six-day period assessed.

Epidermal growth factor (EGF) was induced between two and three days after bulbectomy, i.e., later than LIF. EGF mRNA induction coincided with the end of apoptosis and with onset of progenitor mitotic stimulation.

TGFβ1 and TGFβ3 mRNAs were slightly increasing following bulbectomy, whereas expression of BDNF and TGFβ2 mRNAs remained unchanged during the six-day period assessed (data not shown). Both receptors TGFβ R1 and TGFβ R3 were down-regulated following bulbectomy (data not shown).

Discussion

We analysed gene expression in the olfactory organ using a semi-quantitative RT-PCR approach. The kinetics of mRNA contents in mouse olfactory organ, after normalization onto β-actin mRNA, were highly variable among the eight amplified genes, some of which were induced and others repressed or unaffected.

The data presented should be taken to be indicative of changes in the abundance of mRNAs between samples and only a rough indicator of absolute mRNA levels. Efficiency of RT-PCR can indeed vary at both steps, i.e., RT because of RNAs' secondary structure and PCR because of primers' definition (Estus et al. 1994).

We observed that distinct expression patterns characterized the different mitogenic signals tested. Among all of these, the most interesting were selective inductions of LIF and EGF.

LIF is a multifunctional cytokine that has been shown to exert trophic effects on various types of peripheral and central neurons (Cheng and Patterson 1997; Murphy et al. 1997). *In vitro,* low doses of LIF had been shown to stimulate bromodeoxyuridine incorporation within 24–48 hours in semi-purified olfactory neural progenitors cultures (Satoh and Yoshida 1997a), and in globose basal cells from explanted murine olfactory epithelium (Satoh and Yoshida 1997b). Present induction of LIF mRNA found in the olfactory organ following bulbectomy occurred before the mitotic stimulation of neural precursors, and is thus compatible with the *in vitro* results of Satoh and Yoshida. Present indication for transient lesion-induced surge of LIF signalling in the olfactory organ was still strenghthened by concomitant LIF receptor mRNA up-regulation.

Similarly, sciatic nerve section in adult rat had been shown to involve induction of LIF mRNA after 24 hours (Banner and Patterson 1994) and to increase retrograde transport of LIF along axons of peripheral sensory and motor neurons (Curtis et al. 1994); however LIF receptor mRNA was down-regulated in this paradigm (Banner and Patterson 1994). In the central nervous system, LIF mRNA is also up-regulated rapidly after cortical lesion (Banner et al. 1997). On the other hand, LIF was shown to control the division/differentiation fate of various stem cells in embryonic and adult mammals (Savatier et al. 1996). Olfactory epithelium of adult mammals thus accumulates the two documented types of LIF biological actions.

However, LIF mRNA is short-lived (Derigs and Boswell 1993), so that biological availability of LIF might be too restricted in time to support its mitogenic effect. Yet, LIF might exert this effect indirectly, by inducing EGF responsiveness in basal cells of olfactory epithelium. Our data indeed showed EGF induction just preceding the bulbectomy-induced mitotic stimulation, which strongly suggests that EGF may be a direct mitogenic factor. Consistent with this, is the fact that, in adult mouse olfactory mucosa, EGFR mRNA was detected in all mitotically active basal cells (Krishna et al. 1996), and systemic injection of EGF was shown to activate *in vivo* the transduction pathway usually coupled with its receptor (Ezeh and Farbman 1997). Moreover, interactions between LIF and EGF signalling have been reported (Gadient et al. 1998), so that EGF might relay LIF within olfactory mucosa to trigger a mitotic reaction to lesion-induced apoptosis. This hypothesis is challenged however by reported restriction of EGFR protein to the subpopulation of basal cells that does not react to the present lesion paradigm (horizontal basal cells; Salehi-Ashtiani and Farbman 1996). Cellular localization of LIF,

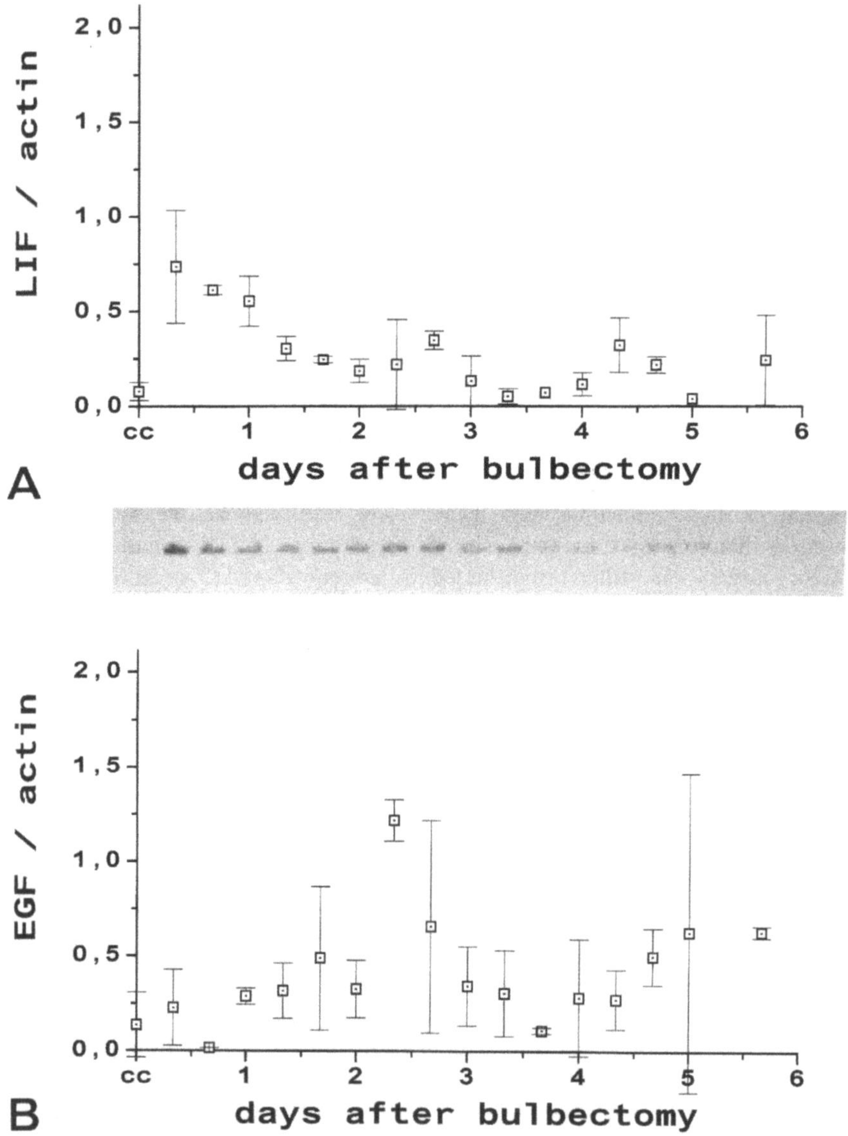

Fig. 2. LIF over β-actin (**A**) and EGF over β-actin (**B**) PCR signal ratios as functions of time after bilateral bulbectomy. Each point is mean ± SEM of RNA samples from three animals, assayed separately but in the same RT and the same PCR experiment. A representative series of RT-PCR signals is given for LIF in **A**, with each band corresponding to one animal and being aligned on corresponding post-lesion time of plot axis

LIFR, and EGF by *in situ* hybridization and/or immunohistochemistry is therefore required to elucidate the respective roles of LIF and EGF in paracrine control of olfactory neurogenesis.

The relevance of LIF to neuronal proliferation should now be tested further with LIF knock-out animals (Bugga et al. 1998).

Other signalling systems assessed were not significantly altered during lesion-induced renewal of olfactory sensory neurons. Therefore. among the cytokines and growth factors that had been shown to stimulate *in vitro* proliferation of olfactory neuronal cells in culture, our study supported only LIF and EGF as physiological mitogens for olfactory neuronal lineage.

Acknowledgements. This study was supported by ARC n° 1377 to E.M.

References

Banner LR, Patterson PH. (1994) Major changes in the expression of the mRNAs for cholinergic differentiation factor/leukemia inhibitory factor and its receptor after injury to adult peripheral nerves and ganglia. Proc Natl Acad Sci USA 91:7109–7113

Banner LR, Moayeri NN, Patterson PH (1997) Leukemia inhibitory factor is expressed in astrocytes following cortical brain injury. Exp Neurol 147:1–9

Bugga L, Gadient RA, Kwan K, Stewart CL, Patterson PH (1998) Analysis of neuronal and glial phenotypes in brains of mice deficient in leukemia inhibitory factor. J Neurobiol 36:509–524

Caggiano M, Kauer JS, Hunter DD (1994) Globose basal cells are neuronal progenitors in the olfactory epithelium: a lineage analysis using replication-incompetent retrovirus. Neuron 13:339–352

Chatelain G, Brun G, Michel D (1995) Screening of homozygous transgenic mice by comparative PCR. Biotechniques 18:959–961

Cheng JG, Patterson PH (1997) LIF is an autocrine factor for sympathetic neurons Mol Cell Neurosci 9:372–380

Crews L, Hunter D (1994) Neurogenesis in the olfactory epithelium Persp Dev Neurobiol 2:151–161

Curtis R, Scherer SS, Somogyi R, Adryan KM, Ip NY, Zhu Y, Lindsay RM, DiStefano PS (1994) Retrograde axonal transport of LIF is increased by peripheral nerve injury; correlation with increased LIF expression in distal nerve. Neuron 12:191–204

Deckner M-L, Frisén J, Verge VMK, Hökfelt T, Risling M (1993) Localization of neurotrophin receptors in olfactory epithelium and bulb. Neuro Report 5:301–304

DeHamer MK, Guevara JL, Hannon K, Olwin BB, Calof AL (1994) Genesis of olfactory receptor neurons in vitro: regulation of progenitor cell divisions by fibroblast growth factors. Neuron 13:1083–1097

Derigs HG, Boswell HS (1993) LIF mRNA expression is transcriptionally regulated in murine bone marrow stromal cells. Leukemia 7:630–634

Estus S, Zacks WJ, Freeman RS, Gruda M, Bravo R, Johnson EM, Jr (1994) Altered gene expression in neurons during programmed cell death: identification of *c-jun* as necessary for neuronal apoptosis. J Cell Biol 6:1717–1727

Ezeh PI, Farbman AI (1997) Functional activation of EGF-R in the olfactory epithelium by TGF-*a* *in vivo.* Soc Neurosci New Orleans, Abstract 458:3

Farbman AI, Bucholtz JA (1996) Transforming growth factor-*a* and other factors stimulate cell division in olfactory epithelium *in vitro.* J Neurobiol 30:267–280

Gadient RA, Lein P, Higgins D, Patterson PH (1998) Effect of leukemia inhibitory factor (LIF) on the morphology and survival of cultured hippocampal neurons and glial cells. Brain Res 798: 140–146

Goldstein BJ, Wolozin BI, Schwob JE (1997) suppresses neurogenesis of a cell line derived from rat olfactory epithelium. J Neurobiol 33:411–428

Gray A, Dull TJ, Ullrich A (1983) Nucleotide sequence of epidermal growth factor cDNA predicts a 128,000-molecular weight protein precursor Nature 303:722–725

Krishna NSR, Little SS, Getchell TV (1996) Epidermal growth factor receptor mRNA and protein are expressed in progenitor cells of olfactory epithelium. J Comp Neurol 373:297–307

MacDonald KPA, Murrell WG, Bartlett PF, Bushell GR, Mackay-Sim A (1996) FGF2 promotes neuronal differentiation in explant cultures of adult and embryonic mouse olfactory epithelium. J Neurosci Res 44:27–39

Mahalik TJ (1996) Apparent apoptotic cell death in the olfactory epithelium of adult rodents: death occurs at different developmental stages. J Comp Neurol 372:457–464

Mahanthappa NK, Schwarting GA (1993) Peptide growth control of olfactory neurogenesis and neuron survival *in vitro*: roles of EGF and TGF-βs. Neuron 10:293-–305

Michel D, Moyse E, Brun G, Jourdan F (1994) Induction of apoptosis in rat olfactory neuroepithelium by synaptic target ablation. NeuroReport 5:1329–1332

Michel D, Chatelain G, Mauduit C, Benahmed M, Brun G (1997) Recent evolutionary acquisition of alternative pre-mRNA splicing and 3′ processing regulations induced by intronic B2 SINE insertion. Nucl Acid Res 25:3228–32234

Monti-Graziadei AG, Graziadei PPC (1979) Studies on neuronal plasticity and regeneration in the olfactory system: morphologic and functional characteristics of the olfactory sensory neurons. In: Meisanic A, Brazier MAB (eds) Neural growth and differentiation. Raven Press, New York, 373–396

Murphy M, Dutton R, Koblar S, Cheema S, Bartlett P (1997) Cytokines which signal through the LIF receptor and their actions in the nervous system. Prog Neurobiol 52:355–378

Salehi-Ashtiani K, Farbman AI (1996) Expression of neu and neu differentiation factor in the olfactory mucosa of the rat. Int J Dev Neurosci 14:801–811

Satoh M, Yoshida T (1997a) Enhancement of neuronal progenitor cell divisions by fibroblast growth factor-2 and leukemia inhibitory factor in organ culture of murine olfactory epithelium. IFO Res Commun 18:13–19

Satoh M, Yoshida T (1997b) Promotion of neurogenesis in mouse olfactory neuronal progenitor cells by leukemia inhibitory factor *in vitro.* Neurosci Lett 225:165–168

Savatier P, Lapillonne H, van Grunsven LA, Rudkin BB, Samarut J (1996) Withdrawal of differentiation inhibitory activity/leukemia inhibitory factor up-regulates D-type cyclins and cyclin-dependent kinase inhibitors in mouse embryonic stem cells. Oncogene 12:309–322

Schwartz-Levey M, Chikaraishi DM, Kauer JS (1991) Characterization of potential precusor population in the mouse olfactory epithelium using immunocytochemistry and autoradiography. J Neurosci 11:3556–3564.

Vawter MP, Basaric-Keys J, Li Y, Lester DS, Lebovics RS, Lesch KP, Kulaga H, Freed WJ, Sunderland T, Wolozin B (1996) Human olfactory neuroepithelial cells: tyrosine phosphorylation and process extension are increased by the combination of IL- 1β, IL-6, NGF, and bFGF. Exp Neurol 142:179–194

Neuroglial Activation in the Nervous System Following Injury: Role of Cytokines

G. W. Kreutzberg

Summary

Current data help to define the physiological role of pro- and anti-inflammatory cytokines like MCSF, IL-6, or TGFβ1 in the injured neural tissue. The combination of knockouts and injury models provides a new approach to gain insight into the cellular and molecular activation cascade that regulates plasticity and regeneration in the lesioned nervous system. The facial nucleus of the brain stem following nerve transection has been used as an experimental model of the cellular reactions. This remote lesion triggers changes primarily in the neuronal transcriptome, e.g., for immediate early genes. This is followed by activation of local microglial cells and reactions of astrocytes. Cytokines are crucial signals in the interaction of the cells in situ.

Introduction and Experimental Model

In an attempt to understand neuronal regeneration, we have been studying for many years the responses of axotomized motoneurons and their cellular environment. As a result of numerous experiments performed mainly in the facial nucleus following the transection of the facial nerve, it became evident that this paradigm is a model for the interaction of neurons and glial cells responding to a challenge by a remote lesion. The key players are the motoneurons, the activated microglial cells and the reactive astrocytes. These three cellular elements constitute a "ménage à trois" under stress (Fig. 1). Depending on the experimental conditions, T-lymphocytes, perivascular cells and vascular cells may act as bystanders in the game. Transection or crush of the facial nerve is performed at the stylomastoid foramen. This standard point guarantees excellent reproducibility and excludes variations in the distance between the lesion and the nerve cell body which could influence the strength of the reaction. The facial nucleus is the largest compact motor nucleus in mammals. It contains motoneurons exclusively, no interneurons. On the unoperated side the intact facial nucleus represents an ideal intra-animal control. The peripheral nerve lesion has no influence on the blood-brain barrier (Kreutzberg 1996).

Patterson/Kordon/Christen (Eds.)
Neuro-Immune Interactions in Neurologic and Psychiatric Disorders

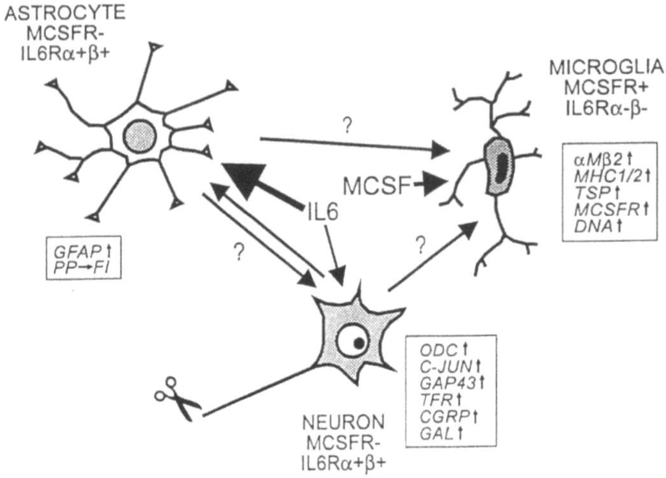

Fig. 1. Ménage à trois. A schematic summary of changes occurring at the level of the injured motoneuron and its glial satellites. In the axotomized neuron, there is a rapid increase in nucleic acid metabolism-associated enzymes (ornithine decarboxylase, ODC), transcription factors (c-JUN), growth-associated proteins (GAP43) and growth factor receptors (transferrin receptor, TFR). Adjacent astrocytes produce growth factors and cytokines (GF/CK), show an increase in GFAP content, and change shape from a protoplasmic to a fibrillary-like type. The third cell type, the microglia, shows enhanced expression of cell adhesion molecules (*αMβ*2-integrin, thrombospondin/TSP), major histocompatibility complex antigens (MHC1,2) and receptors for microglial mitogens (MCSFR). These activated microglia move into direct apposition to the neuronal cell body where they begin to proliferate (DNA). Absence of MCSF leads to a strong reduction in the activation of MCSF receptor-positive microglia but does not affect the receptor-negative neurons and astrocytes. Absence of IL-6 receptor-positive astrocytes and neurons, but also the receptor-negative microglia, indicate an indirect effect. Our aim is to identify the molecular factors, symbolized by the question marks, which mediate this neuroglial cross-talk. Modified from Raivich et al. 1996

The Nerve Cell Body Reaction to Axotomy

Facial nerve axonal injury leads to a well-organized sequence of morphological and metabolic changes in and around the affected neuron which appear to play an important role in preventing further damage and enabling repair of the injured neural tissue. The cell bodies of axotomized neurons show rapid cytoplasmic swelling, eccentric nuclei and an increase and redistribution of ribosomes that lead to the light-microscopic dissolution of Nissl bodies, so-called chromatolysis. At the molecular level, there is an early induction in RNA metabolism and associated enzymes, ornithine decarboxylase and transglutaminase. This is followed by the expression of the nuclear transcription factors c-jun, junB, junD and TIS11, and increased synthesis of growth associated proteins (GAPs) such as the growth cone Ca-modulating protein GAP-43/neuromodulin. Injured neurons also show an increase in neuronal energy metabolism, glucose and iron uptake, and the induction of growth factors, growth factor receptors and other axotomy-associated signaling molecules (Kreutzberg et al. 1997; Schwaiger et al. 1998).

Such fundamental changes occur in the CNS only in axotomized motoneurons, which are well known to grow new axonal sprouts and reinnervate the periphery. Other CNS neurons, e.g., in the red nucleus, Clarke's nucleus or other areas, show only very weak upregulation of transcription factors, immediate early genes, or GAPs. Hence the changes in the transcriptome of the motoneuron have been interpreted as a genetically established regeneration program of neurons. These neuronal changes are accompanied by alterations in the satellite glial cells.

Activation of Microglial Cells

The microglial cell is a member of the monocyte family and forms the resident macrophage of the CNS tissue (Perry and Gordon 1988). Precursors enter the neural parenchyme during embryogenesis and are recognized as ameboid microglia. They develop into typical ramified forms after birth and settle in non-overlapping territories as resting microglial cells. They account for approximately 20% of the total neuroglial population. In various neuropathological states microglia play a pivotal role as the first line of defense. The fact that a deficient blood-brain barrier also allows the entry of blood-borne monocytes/macrophages into the brain has strongly clouded the correct recognition of microglial activation that can lead to the intrinsic formation of brain macrophages.

To study microglial experimentally in vivo, certain prerequisites should be maintained. The blood-brain barrier should remain intact so that contamination by blood-borne cellular elements, e.g., monocytes, as occurs in a penetrating lesion, can be excluded. The postaxotomy facial nucleus offers such a situation.

Microglial cells adjacent to the chromatolytic motoneurons in the experimental facial nucleus show a rapid increase in cell adhesion molecules, receptors for microglial mitogens and cytoskeletal proteins. Within the first 48 hours after lesion, these activated microglia adhere to the perikaryal surface of injured neurons, displace synaptic input and proliferate vigorously, with a 4 to 6-fold increase in microglial cell number seven days after axotomy (Blinzinger and Kreutzberg 1968).

Activated microglial cells become immunocompetent and are MHC class 1 and class 2-positive. They express the amyloid precursor protein, (APP). The complement receptor CR3bi ($\alpha M\beta 2$-integrin) is quickly upregulated. The mitotic activity depends on the colony stimulating factors MCSF and GMCSF and the appropriate receptors. Molecules discussed as signals in the activation process of microglia are cytokines such as IL-1, IL-2, IL-6, and TGFβ1.

An important role could also be attributed to the unique potassium channel of microglia. Brain macrophages of microglial origin have a strong respiratory burst activity, meaning that they produce oxygen radicals. They also possess cathepsin B and L and thus are potentially cytotoxic (Kreutzberg 1996). This finding confirms the view that microglia are highly reactive, mobile and multifunctional immune cells of the CNS and that they can play a universal role in the defense of the neural parenchyma. They belong to the mononuclear phagocyte system and form the resident macrophages in the brain tissue, the spinal cord and the retina. Their function in the normal neural parenchyma is unknown. However, in various pathologies they form a most reactive sensor to threats to the nervous system. Within a few hours they exhibit an activation program that we have studied in seven different experimental paradigms, e.g., following nerve section, direct brain trauma, toxic lesion, spreading depression, ischemic lesion, fiber degeneration, and autoimmune diseases. These studies established the view that activated microglial cells display a repertoire of molecular regulation not specifically related to the pathogenetic event.

In vitro studies have identified a growing number of mitogens for microglia-derived macrophages. However, most of these studies were performed on rounded macrophages, which show numerous differences compared to ramified microglia, the main proliferating, monocyte-related cell in the developing or lesioned central nervous system. It is now possible to investigate stable cell cultures of ramified microglia. They are obtained using a 10-day co-culture on top of a confluent astrocyte monolayer. The addition of the classical macrophage mitogens, macrophage colony-stimulating factor (MCSF), granulocyte-macrophage colony-stimulating factor (GMCSF) and interleukin-3 (IL3), strongly stimulated the proliferation of ramified microglia. Of all the other cytokines tested (IL-1, IL-2, IL-6. IL-10, IFNγ, LIF and TNFα), only IL-1 strongly enhanced proliferation. This effect, however, was indirect and neutralized by antibodies against MCSF and GMCSF. IL-2, IL-4, IL-10, TNFα and IFNγ inhibited microglial proliferation. IL-6 and LIF had a slight, but nonsignificant, stimulatory influence. Surprisingly, only the classical macrophage-mitogens (MCSF, GMCSF, IL-3) promoted proliferation directly (Kloss et al. 1997).

Microglia are considered to be the most important immune cells of the brain tissue. Therefore, the expression of immunoregulatory molecules known for their role in antigen presentation is especially intriguing. The immune response is characterised by the binding of antigen to glycoproteins of the major histocompatibility complex MHC class 1 or MHC class 2 supported by immunoaccessory molecules, e.g., B7-1, B7-2, ICAM-1.

Ongoing studies in our lab (Bohatschek et al. 1998) have shown that in the postaxotomy facial nucleus of mice a selective upregulation of MHC 1 and B7-2 occurs on microglia. In contrast to the rat, in mice some 10–20% of motoneurons do not survive the axotomy stress and undergo cell death after two weeks. Thus phagocytotic microglia are recruited to the neuronal graves (Fig. 2). They express particularly high levels of MHC I, which also applies to the perivascular cells. In fact the infiltration of T-helper and suppressor lymphocytes in the mouse demonstrates a highly localised immune surveillance for challenged neural tissue (Hickey et al. 1991; Raivich et al. 1998).

In our model we observed a rapid influx of CD3-, CD11a (αL, LFAα)- and CD44-immunoreactive T-cells into the mouse facial nucleus, with a first,

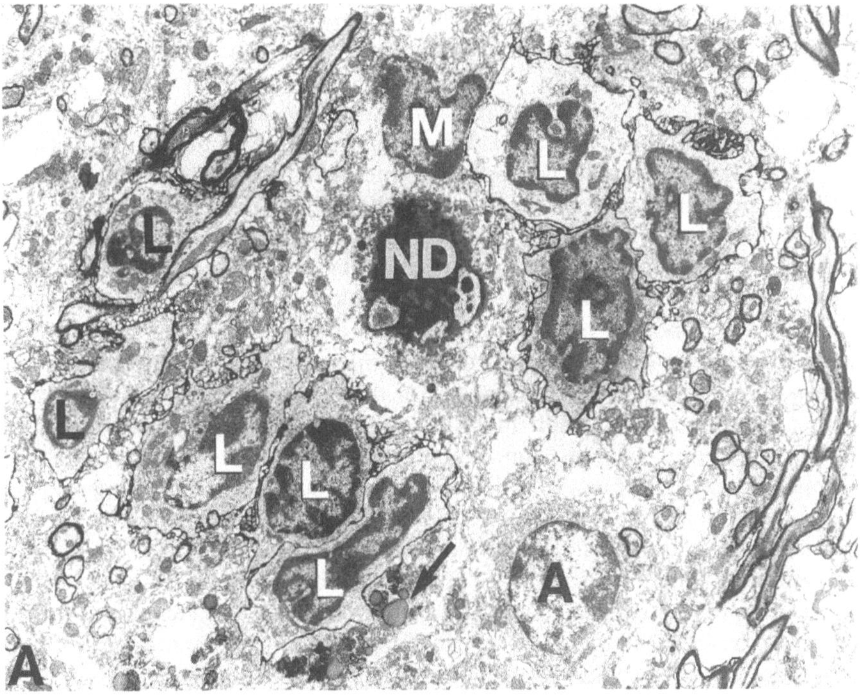

Fig. 2. "Neuronal grave" 14 days after mouse facial nerve axotomy. CD11a immunostaining of a cellular aggregate, consisting of the debris of a degenerating neuron (ND) phagocytosed by microglial cells (M and arrow), an astrocyte (A), and the lymphocytes (L). The latter show positive CD11a immunoreaction on the ruffled plasma membrane. Magnification 5400x. (Raivich et al. 1998)

low-level plateau two to four days after injury and a second, much stronger increase at 14 days. These T-cells frequently formed aggregates and exhibited typical cleaved lymphocyte nuclei at the EM level. Immunohistochemical colocalization with thrombospondin, a marker for phagocytotic microglia, revealed aggregation of the T-cells around microglia removing neuronal debris (Fig. 2). The massive influx of lymphocytes at day 14 was also accompanied by the synthesis of mRNA encoding IL1β, TNFα, and IFN-γ. There was no infiltration by the neutrophil granulocytes, and the intravenous injection of horseradish peroxidase also showed an intact blood-brain barrier. However, mice with severe combined immunodeficiency (SCID), which lack differentiated T- and B-cells, still exhibited infiltration with CD11a-positive cells. These CD11a-positive cells also aggregated around phagocytotic microglial nodules (Raivich et al. 1998) The coordinated appearance of antigen-presenting microglia and perivascular cells and the appropriate lymphocytes stresses the concept of an immune surveillance in the CNS (Wekerle et al. 1986).

In addition to immune regulatory molecules, the induction of new cell adhesion molecules takes center stage in the interaction of microglia with their cellular environment after injury. Intercellular adhesion molecule 1 (ICAM-1) is a widely expressed adhesion protein, which is a member of the immunoglobulin superfamily. ICAM-1 is a ligand for the lymphocyte function-associated-1 antigen (LFA-1/$\alpha L\beta 2$-integrin) and plays an important role in the extravasation of leukocytes during inflammation of neural and peripheral tissue. In the injured CNS, activated microglia also express high levels of $\alpha M\beta 2$-integrin, the second leukocyte receptor for ICAM-1. Transection of the facial nerve led to a strong and selective upregulation of ICAM-1 on the activated microglia and its expression on lymphocytes invading the injured nervous tissue (Werner et a. 1998). This finding suggests the involvement of ICAM-1 in lymphocyte infiltration and in mediating cell-to-cell contacts between endothelial cells, lymphocytes and microglia during the cellular response to injury and the ensuing repair of the central nervous system.

Yet another cell adhesion molecule is CD44, which plays an important role in cell movement, adhesion, homing and settlement of cells outside the nervous system. However, CD44 is also rapidly upregulated and appears on the surface of axotomized motoneurons, suggesting a possible role in the guidance and adhesion of perineuronal microglia (Jones et al. 1997).

Neuronal adhesion of activated microglia plays a central role in the lifting of synaptic input from the surface of the cell body and stem dendrites of the axotomized and regenerating motoneuron, a deafferentation process known as "synaptic stripping" (Blinzinger and Kreutzberg 1968).

Microglial activation has been traditionally difficult to study mainly because there are no cell type-markers. All known markers for microglia are microglia-specific only in the sense that they do not label other glia or neurons. Widely used marker molecules include lectins as well as a growing number of monoclonal antibodies. Yet, all monoclonal antibodies cross-react with macrophages from other organs or monocytes in the blood. It was not until the advent of an experimental model system such as the facial nerve and nucleus that stages of microglial activation could be studied in a systematic manner. Recently, a so-called peripheral benzodiazepine binding site to PK11195 has been found to be a characteristic of activated microglial cells. Such binding can be seen upregulated in the facial nucleus after experimental and clinical lesions. Since PK11195 can be labeled by means of the isotope [^{11}C], the radioactive ligand becomes visible in the living organisms using positron emission tomography. Such a methodology is now developed and will open new vistas for early diagnosis in lesions of the human CNS (Banati et al. 1997).

Reactive Changes in Astrocytes

Following indirect injury, e.g., in the facial model, supraneuronal astrocytes undergo a fast induction of cytokines and extracellular matrix proteins, and a massive increase in the cytoskeletal glial fibrillary acidic protein (GFAP), a reactive key marker of astrocytes. They change shape from a protoplasmic to a stellar, fibrillar cell type. These astrocytes gradually displace microglia from the neuronal surface and enwrap injured neurons with astrocytic lamellae (Graeber and Kreutzberg 1990; Kreutzberg et al. 1997). Like microglia, astrocytes display a graded response to injury. Their reaction follows the severity of the pathological event. Upregulation of GFAP, insulin-like growth factor 1 (IGF-1) and platelet-derived growth factor (PDGF) are frequent companions of astroglia reactivity. The signaling molecules that lead to this cellular response are just beginning to be understood. Our recent work, combining data on the expression of these signaling molecules, their receptors and the effect of cytokine knockouts, has allowed the identification of MCSF and IL-6 as key factors in microglial and astrocyte activation. The role of MCSF in microglial activation has already been described. IL-6 is a proinflammatory cytokine that shares receptor sites with the leukemia inhibitory factor (LIF) and CNTF. Overexpression of IL-6 leads to disturbances of the blood-brain barrier, gliosis and neuronal damage. Induction of IL-6 is common to all tissue damage in the brain and is seen in trauma, ischemia, inflammation, and even in neurodegenerative diseases. IL-6 receptors are localized on neurons and astrocytes, but not on microglia, and this expression pattern does not change after injury.

The absence of IL-6 following transgenic deletion leads to striking changes in the pattern of neuroglial activation (Klein et al. 1997). In the axotomized facial motor nucleus model, all three major cell types – neurons, astrocytes and microglia – are affected. Compared to normal, wildtype control mice, the IL-6-deficient animals show a much smaller, and sometimes completely absent, posttraumatic induction of GFAP-positive astrocytes. More moderate effects are observed on microglial proliferation. On the other hand, injured neurons respond by a late and strong increase in galanin, a neuropeptide expressed during motoneuron regeneration.

The current data suggest a hierarchy in the cellular action of IL-6, with reactive astrocytes being the primary target. Astrocytes express high levels of IL-6 receptors and they are particularly strongly affected by IL-6 deficiency. The effects on astrocytes also closely follow the onset of IL-6 expression following injury. The absence of microglial receptors clearly suggests that the IL-6 action on microglia is indirect. The late onset in the neuronal expression of galanin could also point to an indirect effect, emphasizing the role of reactive astrocytes in the overall orchestration of cellular response in the injured nervous system.

Transforming growth factor-beta 1 (TGFβ1) is an anti-inflammatory cytokine with potent neurotrophic and immunosuppressive properties. TGFβ1 is already present at low levels in the normal central nervous system, but the expression of this cytokine is strongly upregulated in almost all forms of brain pathology, quite similar to the regulation pattern for IL-6 (Kiefer et al. 1993, 1995). On cultured astrocytes, TGFβ1 leads to a strong increase in the synthesis of the cytoskeletal protein GFAP and extracellular matrix (laminin, fibronectin) but inhibits their proliferation. In TGFβ1-deficient animals an enhanced astroglial reactivity is seen in the facial model.

Conclusion

The remote lesion model of the facial nucleus has revealed cellular and molecular interactions between lesioned neurons and glial cells in vivo. The neuron as the primary address of an injury signal initiates activation of microglial cells and hypertrophic reactions of astrocytes. In this complex network a state of alert is achieved which comprises immune mechanisms, neurotrophic and regenerative action and, in cases of increasing threat, also cytotoxic and phagocytotic mechanisms. The network is orchestrated by cytokines and other modulatory factors and their receptors.

References

Banati RB, Myers R, Kreutzberg GW (1997) PK ("peripheral benzodiazepine")-binding sites in the CNS indicate early and discrete brain lesions: microautoradiographic detection of [^{3}H]PK11195 binding to activated microglia. J Neurocytol 26:77–82

Blinzinger K, Kreutzberg G (1968) Displacement of synaptic terminals from regenerating motoneurons by microglial cells. Zeitschr Zellforsch 85:145–157

Bohatschek M, Jones LL, Kreutzberg GW, Raivich G (1998) Expression of immunoregulatory molecules MHC1, MHC2 and B7-2 in the axotomized mouse facial motor nucleus. Clin Neuropathol No5 Vol 17:286

Graeber MB, Kreutzberg GW (1990) Astrocytic reactions accompanying motor neuron regeneration. In: Seil J (ed) Advances in neural regeneration research. New York, Wiley-Liss Inc., pp 215–224

Hickey WF, Hsu BL, Kimura H (1991) T-lymphocyte entry into the central nervous system. J Neurosci Res 28:254–260

Jones LL, Kreutzberg GW, Raivich G (1997) Regulation of CD44 in the regenerating mouse facial motor nucleus. Eur J Neurosci 9: 1854–1863

Kiefer R, Lindholm D, Kreutzberg GW (1993) Interleukin-6 and transforming growth factor beta-1 mRNAs are induced in rat facial nucleus following motoneuron axotomy. Eur J Neurosci 5:775–781

Klein MA, Möller JC, Jones LL, Bluethmann H, Kreutzberg GW, Raivich G (1997) Impaired neuroglial activation in interleukin-6 deficient mice. Glia 19:227–233

Kloss, CUA, Kreutzberg GW, Raivich G (1997) Proliferation of ramified microglia on an astrocyte monolayer: characterization of stimulatory and inhibitory cytokines. J Neurosci Res 49:248–254

Kreutzberg GW (1996) Microglia: a sensor for pathological events in the CNS. Trends Neurosci 19:312–318

Kreutzberg GW, Blakemore WF, Graeber MB (1997) Cellular pathology of the central nervous system. In: Graham DI and Lantos PL (eds) Greenfield's neuropatholog. 6th Ed. London, Oxford University Press, pp 85–156

Perry VH, Gordon S (1988) Macrophages and microglia in the nervous system. Trends Neurosci 11:273–277

Raivich G, Bluethmann H, Kreutzberg GW (1996) Signaling molecules and neuroglial activation in the injured central nervous system. Keio J Med 45:239–247

Raivich G, Jones LL, Kloss CUA, Werner A, Neumann H, Kreutzberg GW (1998) Immune surveillance in the injured nervous system: T-Lymphocytes invade the axotomized mouse facial motor nucleus and aggregate around sites of neuronal degeneration. J Neurosci 18(15):5804–5816

Schwaiger FW, Hager G, Raivich G, Kreutzberg GW (1998) Cellular activation in neuroregeneration. Prog Brain Res 117:197–210

Wekerle H, Linington C, Lassmann H, Meyermann R (1986) Cellular immune reactivity within the CNS. Trends Neurosci 9:271–277

Werner A, Kloss CUA, Walter J, Kreutzberg GW, Raivich G (1998) Intercellular adhesion molecule-1 (ICAM-1)in the mouse facial motor nucleus after axonal injury and during regeneration. J Neurocytol 27:219–232

Sickness Behavior: A Neuroimmune-Based Response to Infectious Disease

R. Dantzer, A. Aubert, R.M. Bluthe, J.P. Konsman, S. Laye, P. Parnet, and K.W. Kelley

Summary

Sickness behavior refers to a coordinated set of behavioral changes that develop in sick individuals during the course of an infection. These changes are due to the effects in the brain of proinflammatory cytokines such as interleukin-1 (IL-1) and tumor necrosis factor alpha (TNFα) and represent the expression of a well-organized central motivational state. Based on the results of pharmacological and biochemical experiments, it is now apparent that sickness behavior is mediated by cytokines that are temporarily synthesized and released in the brain during infection. These cytokines act on brain receptors that are identical to those characterized on immune cells. Primary afferent nerves represent the main communication pathway between peripheral and central cytokines. The existence of a brain cytokine compartment that is inducible by peripheral immune stimuli and is associated with sickness behavior has important pathophysiological implications.

The Concept of Sickness Behavior

Non-specific symptoms of infection and inflammation include fever and profound physiological and behavioral changes. Sick individuals experience weakness, malaise, listlessness and inability to concentrate. They become depressed and lethargic, show little interest in their surroundings and stop eating and drinking. This constellation of non-specific symptoms is collectively referred to as "sickness behavior." Due to their commonality, sickness symptoms are frequently ignored by physicians. They are considered an uncomfortable, but rather banal, component of the pathogen-induced debilitative process.

A few years ago, Hart (1988) proposed that the behavioral symptoms of sickness represent, together with the fever response, a highly organized strategy of the organism to fight infection. This view was built on the already recognized role of fever in the host response to pathogens (Kluger 1979). In physiological terms, fever corresponds to a new homeostatic state that is characterized by a raised set-point of body temperature regulation. A feverish individual feels cold at usual thermoneutral environments. Therefore the

Patterson/Kordon/Christen (Eds.)
Neuro-Immune Interactions in Neurologic and Psychiatric Disorders

individual not only seeks warmer temperatures but also tries to enhance heat production (increased thermogenesis) and reduce heat loss (decreased thermolysis). The higher body temperature that is achieved during fever stimulates proliferation of immune cells and is unfavorable for the growth of many bacterial and viral pathogens. In addition, the reduction of zinc and iron plasma levels that occurs during fever decreases the availabililty of these vital elements for growth and multiplication of microorganisms. The adaptive nature of the fever response is apparent from studies showing that organisms infected with a bacteria or virus and unable to mount an appropriate fever response because they are kept in a cold environment or are treated with an antipyretic drug have a lower survival rate than organisms that develop a normal fever (Kluger 1979).

The amount of energy that is required to increase body temperature during the febrile process is quite high since, in human beings, metabolic rate needs to be increased by 13% for a rise of 1 °C in body temperature. Because of this high metabolic cost of fever, there is little room for activities other than those favoring heat production (e.g., shivering) and minimizing thermal losses (e.g., rest, curl up posture, piloerection).

The molecular signals that mediate the local immune reaction are also responsible for ensuring the necessary synchrony between metabolic, physiological and behavioral components of the systemic response to infection. These signals are proinflammatory cytokines, such as interleukin-1α and interleukin-1β (IL-1), interleukin-6 (IL-6), tumor necrosis factor α (TNF-α and interferons (IFNs), which are released by activated monocytes and macrophages during the course of an infection (Baumann and Gauldie 1994).

The objective of the present chapter is to summarize the results that have been obtained recently concerning the mechanisms by which peripherally released IL-1, considered here as a prototypical proinflammatory cytokine (Dinarello 1996), induces sickness behavior.

Proinflammatory Cytokines Induce Sickness Behavior

The first evidence in favor of a possible role of cytokines in sickness behavior came from clinical trials with purified or recombinant cytokines in the treatment of intractable viral diseases and cancer. Patients injected systemically with these molecules were observed to develop not only flu-like symptoms but also, on repetition of injections, acute psychotic episodes characterized by depression or excitation. Fortunately, these symptoms spontaneously regressed on cessation of treatment. Toxicological studies carried out on laboratory animals confirmed the neurotropic activity of recombinant cytokines such as IL-1β and TNF-α. Animals injected acutely or chronically with these molecules usually appear lethargic, anorexic and withdrawn from their environment (Kent et al. 1992b). The same effect is observed when animals are injected with the active fragment of endotoxin, lipopolysaccharide (LPS), which induces the synthesis and release of proinflammatory cytokines by monocytes and macrophages.

More objective studies of the sickness-inducing properties of cytokines have been based on the observation of changes in locomotor activity, social activities and food intake (Dantzer et al. 1993). Adult animals presented with juveniles of the same species spontaneously investigate these social stimuli. This investigatory behavior remains at a high level as long as unfamiliar juveniles are presented. Systemic and central administration of LPS, IL-1β and TNF-α to the adult animal decreases the duration of social investigation (Bluthé et al. 1992, 1994a). These effects are slow to appear since they do not develop before two to four hours after injection. They last for four to six hours and recovery is usually complete by 24 hours. Systemic and central administration of LPS, IL-1 and TNF-α also consistently suppress feeding and drinking. This effect has been observed using various measurements of food and water intake and *ad libitum* as well as deprived conditions (Kent et al. 1995; Plata-Salaman 1996).

Sickness Behavior is the Expression of a Central Motivational State

As mentioned above, sickness behavior is usually viewed by physicians as the result of the debilitation and physical weakness that inevitably occur in an organism whose total resources are engaged in a defensive process against pathogens. The association of sickness behavior with fever implies that sickness behavior is the expression of a highly organized strategy that is critical to the survival of the organism. If this is the case, then it follows that sick individuals should be able to reorganize their behavior, depending on its consequences and the internal and external constraints they are exposed to. This flexibility is characteristic of what psychologists call motivated behavior. Motivation can be defined as a central state that reorganizes perception and action. In order to escape a potential threat, a fearful individual must be attentive to everything that can occur in his environment so as to be able to exhibit at the right time the most appropriate defensive behavioral pattern that is available in his behavioral repertoire. Therefore, a motivational state does not trigger an inflexible behavioral pattern. On the contrary, by uncoupling action from stimulus conditions it enables the individual to select the appropriate strategy depending on the eliciting situation.

The first evidence that sickness behavior is the expression of a motivational state rather than a consequence of weakness was provided by Miller (1964). Rats placed in a Skinner box in which they had to press a lever for getting food, water or intracranial self stimulation decreased their response rate in response to endotoxin. However, when they were submitted to forced exercise by being placed in a rotating drum that they could stop for brief periods by pressing a lever, they increased their response rate in response to endotoxin treatment.

The mere fact that endotoxin treatment can decrease or increase behavioral output depending on its consequences gives strong support to the motivational interpretation of the behavioral effects of such a treatment. To further assess the validity of the motivational interpretation of sickness behavior, Aubert et al. (1995 a, 1997 a, b manuscript in preparation) engaged in an extensive series of experiments. In one experiment (Aubert and Dantzer, manuscript in preparation), rats were trained to work for food on a progressive ratio schedule. In the schedule that was chosen for this particular experiment, rats had to press the lever once to get the first pellet, twice to get the second pellet, four times to get the third pellet, eight times for the fourth pellet, and so on. This means that food became more and more difficult to obtain with time. Normally, the animal stops pressing the lever when the ratio becomes too high in comparison to its motivation for food. If food motivation is decreased, the achieved ratio is smaller whereas if food motivation increases, the achieved ratio is larger. When well-trained rats had to work for food in a cold environment (4–5 °C), some increased their response rate whereas others responded less. These changes correspond to two different thermoregulatory strategies, consisting of either increasing energy intake to produce more heat or stopping lever pressing and adopting a curled posture to decrease energy expenditure and minimize thermal losses. The important result is that when rats were injected with a small dose of IL-1β directly into the lateral ventricle of the brain, they responded to this immune signal in exactly the same way as when they were exposed to cold, e.g., those that enhanced their response rate in the cold enhanced their response rate in reponse to IL-1β and vice versa.

Another important characteristic of a motivational state is that it competes with other motivational states for behavioral output. It is normally not possible to search for food and court a potential sexual partner at the same time, since the behavioral patterns of foraging and courtship are not compatible with each other. The normal expression of behavior therefore requires a hierarchical structure of motivational states that is updated contingent on urgencies that occur in the internal and external milieux. When an infection occurs, the sick individual is at a life or death juncture and its physiology and behavior must be altered to overcome the disease. However, this is a relatively long-term process that needs to make room to more urgent needs when necessary. To make an analogy, it is obvious that if a sick person lying in his bed hears a fire alarm ringing in his house and sees flames and smoke coming out of the basement, he will momentarily overcome his sickness behavior to try to escape danger. In motivational terms, fear competes with sickness and, in behavioral terms, fear-motivated behavior supplants sickness behavior. An example of this competition between sickness and other motivational states is provided by the observation that the depressing effects of LPS on the behavior of lactating mice were abrogated when LPS-treated mothers had to retrieve their pups after they had been removed from the nest and scattered throughout the home cage (Aubert et al. 1997 a), Additionally, LPS-treated mothers also engaged in nest building when the motivation to build a nest was increased by placing the mother with her litter at an ambient temperature of 6 °C, whereas nest building did not occur when the animals were tested at 22 °C.

Cytokines and Cytokine Receptors Are Expressed in the Brain

If sickness behavior is the expression of a central motivational state, it must be organized in the brain and triggered by the brain actions of peripheral immune signals. The mechanisms by which peripheral immune signals act on the brain have been elucidated in recent years. Equilibrium binding of recombinant human ^{125}I-IL-1α by brain membrane homogenates and brain slices has revealed the presence of binding sites for IL-1 in the mouse but not the rat brain (Haour et al. 1990; Takao et al. 1990). These binding sites are located almost exclusively in the dentate gyrus of the hippocampus and the choroid plexus. They are also found in the anterior pituitary. *In situ* hybridization identified the type I IL-1 receptor mRNA in several brain structures including the anterior olfactory nucleus, thalamus, hypothalamus, amygdala and cerebellum (Cunningham et al. 1992). Double labeling revealed that the signal for the type I IL-1 receptor is mainly localized in neuronal cells, endothelial cells and epithelial cells of the choroid plexus and ventricles. RT-PCR techniques have revealed the existence of both type I and type II IL-1 receptor mRNAs in various brain regions in the mouse (Parnet et al. 1994). Neuronal cell lines derived from chromaffin cells express only the type I IL-1 receptor. Immunocytochemistry techniques with rat monoclonal antibodies directed against the different types of IL-1 receptors in the mouse system have confirmed that both the type I and type II IL-1 receptors are present in the hippocampus, cerebellum and choroid plexus, whereas only the type II IL-1 receptor is found in the paraventricular nucleus of the hypothalamus (French et al. 1999). In the anterior pituitary, both types of IL-1 receptors are present only on those pituitary cells that synthesize and release growth hormone (Parnet et al. 1993; French et al. 1996).

The expression of cytokine receptors in the brain is associated with the presence of brain cytokines. IL-1 is expressed in the brain in two pools: a constitutive form that appears to be mainly present in neurons, and an inducible form that is expressed mainly in brain macrophages and microglial cells (Hopkins and Rothwell 1995). Peripheral immune activation produced, for instance, by systemic administration of LPS induces the expression of IL-1β transcripts in the brain (Layé et al. 1994). The same treatment induces the expression of immunoreactive IL-1 first in circumventricular organs and then in the brain parenchyma where it is produced mainly by microglial cells (Van Dam et al. 1992; Buttini and Boddeke 1995; Konsman et al. 1999). Constitutive expression of the endogenous receptor antagonist of IL-1 (IL-1ra) has been described in the rat brain. Its mRNA is induced in the brain in response to LPS and it peaks later than IL-1 (Licinio et al. 1991).

Mechanisms of the Behavioral Effects of Cytokines

Since cytokines are released locally in the body during the course of infection and inflammation, the question of how these molecules act on the brain has been the object of much debate during the last decades. Cytokines are relatively big proteins (about 150 amino acids for IL-1α and IL-1β) that are hydrophilic and therefore cannot cross the blood-brain barrier. For this reason, they are believed to act on those brain sites, that lack a blood-brain barrier and are known as circumventricular organs because of their spatial location close to the brain ventricles. There, cytokines would trigger the synthesis and release of prostaglandins of the E series that would freely diffuse to the target neuronal cells. This mode of action has been proposed to account for both the corticotropic and pyrogenic effects of cytokines (Katsuura et al. 1990; Stitt 1985).

The problem with this hypothesis is that it does not account for the previously mentioned observation that cytokines are present with their receptors in the brain and their local expression is modulated by peripheral immune stimuli. There are several reasons to propose that central cytokines are responsible for the behavioral effects of peripherally released cytokines. The most obvious one is the fact that much lower doses of IL-1β or TNF-α are needed, when injected directly into the lateral ventricle of the brain or specific brain structures, to depress feeding behavior and social exploration than the doses required when the same cytokine is injected at the periphery (Kent et al. 1992b). The ratio is usually 1 to 100 or 1000. Another finding favoring the possibility of a central site of action is the demonstration of the blockade of the behavioral effects of peripherally injected IL-1β by a central injection of the specific antagonist of IL-1 receptors, IL-1ra (Kent et al. 1992a). In this experiment, rats were pretreated icv with a dose of IL-1ra sufficient to block the depressing effects of icv-injected IL-1β on social exploration and food-motivated behavior. This pretreatment completely abrogated the reduction of social exploration induced by ip administration of IL-1β and attenuated the decrease in response rate that occured in IL-1β-treated rats trained to get their food by pressing a lever in a Skinner box. These results indicate that IL-1 administered at the periphery acts in the brain via classical IL-1 receptors but they do not help to determine where the centrally acting IL-1 comes from, i.e., whether it is imported from the periphery or produced locally in the brain. The demonstration that peripheral administration of LPS induces the expression of cytokines in the brain (Gatti and Bartfai 1993; Layé et al. 1994) is strongly suggestive of the second possibility.

Based on these findings, Kent et al. (1992b) proposed that cytokines released at the periphery activate primary sensory neurons and that this neuronal activation conveys the peripheral immune message to the brain, where it results in the synthesis and release of a functional pool of cytokines. The first evidence in favor of this hypothesis was provided by Wan et al. (1994), who investigated the distribution of Fos expression in the brain in rats injected intraperitoneally with LPS. Fos-immunoreactive neurons were identified in the primary projection area of the afferent branches of the vagus nerve, represented by the nucleus tractus solitarius, and secondary projection areas such as the parabrachial nucleus, the paraventricular nucleus and the supraoptic nucleus of the hypothalamus. Furthermore, transection of the vagus nerve at the subdiaphragmatic level, which eliminates afferent neurons orginating from the liver and the gastro-intestinal tract, abrogated LPS-induced Fos immunoreactivity in these brain areas.

The functional nature of this vagal communication pathway between the immune system and the brain was evidenced by the demonstration that section of the vagus nerve abrogated LPS-induced hyperalgesia in rats (Watkins et al. 1994a) as well as LPS- and IL-1β induced decreases in social exploration (Bluthé et al. 1994b) and food-motivated behavior (Bret-Dibat et al. 1995) in rats and mice. The vagus nerve transmits to the brain not only the signals that are responsible for the behavioral effects of cytokines but also those that are responsible for fever and pituitary-adrenal activation, since vagotomy has been shown to block these two last responses to LPS (Fleshner et al. 1995; Watkins et al. 1995). In accordance with such a wide range of effects of vagotomy on the brain actions of peripheral immune stimuli, this surgical procedure was found to result in the abrogation of the induction of IL-1β in the brain at the mRNA and protein levels in response to peripherally injected LPS (Layé et al. 1995).

Although all these findings are strikingly suggestive of a neural pathway mediating the effects of peripheral cytokines in the brain, other interpretations are possible. In particular, section of the vagus nerve might alter in a non-specific way the sensitivity of circumventricular organs to circulating cytokines because of the close anatomo-functional relationships between one of these circumventricular organs, the area postrema, and the primary projection area of the vagus nerve, the nucleus tractus solitarius. Another possibility is that the neural reorganization that takes place in the projection areas of the vagus nerve decreases the sensitivity of brain cell targets to cytokines. If the consequences of vagotomy on the behavioral effects of cytokines are due to such mechanisms, vagotomized animals should be less sensitive to the effects of cytokines that are administered by routes of injection other than the intraperitoneal route. However, section of the vagus nerve did not attenuate the behavioral effects of IL-1β when this cytokine was injected into the lateral cerebral ventricle of the brain (Bluthé et al. 1996a) and by the subcutaneous or intravenous route (Bluthé et al. 1996b). These results are important because they confirm the role of vagal afferent nerves in the transmission of the immune message from the periphery to the brain and they show that the vagus nerve conveys specific information concerning cytokines injected into the abdominal cavity. The mechanisms that are responsible for the transformation of the immune message in a neural message at the periphery and the transduction of this neural message back into an immune message in the central nervous system still need to be worked out (Dantzer 1994a), but they provide a fascinating example of communication between the brain and the periphery.

Opposition of Behavioral Effects of Ctytokines by Cryogens

At the same time that they affect brain functions, cytokines trigger molecular mechanisms that are part of a feedback loop regulating the neurotropic effects of cytokines and promoting recovery. Since these molecules have originally been identified on the basis of their ability to oppose the pyrogenic effects of cytokines, they have been named endogenous antipyrogens or cryogens (Kluger 1991).

The neuropeptide vasopressin (VP) is a key factor in regulation of water metabolism. It is produced by magnocellular hypothalamic neurons and accumulates in the terminals of these neurons in the posterior pituitary. It is released in the general circulation and acts as an antidiuretic hormone in conditions of water deprivation and in response to fever. VP can also act as a neurotransmitter in the brain: it is present in particular in neurons whose cell bodies are located in the bed nucleus of the stria terminalis (BNST) and the terminals project to the lateral septum. This vasopressinergic pathway is highly sensitive to circulating androgens in rodents. Castration leads to a dramatic decrease in the content of VP mRNA in the BNST neuronal cell bodies and to a reduction in immunoreactive VP in the terminal areas of the septum. This androgen-dependent pathway is also activated during fever.

Central administration of VP attenuated the depressing effects of centrally injected IL-1β on social exploration. Conversely, central injection of an antagonist of vasopressin receptors, which has no biological activity on its own but prevents endogenously released vasopressin to reach its receptors, sensitized rats to the behavioral effects of IL-1. These last results are important since they suggest that endogenous VP plays a physiological modulatory role in the behavioral effects of IL-1. To determine whether this phenomenon is mediated by an androgen-dependent or independent vasopressinergic pathway, castrated male rats were compared to intact male rats in their sensitivity to the modulatory role of the VP receptor antagonist. Castration by itself potentiated the depressing effects of IL-1β on social exploration. Central administration of VP was more effective in attenuating the behavioral effects of IL-1 in castrated than in intact male rats and, conversely, icv administration of the VP receptor antagonist was no longer active in potentiating the behavioral effects of IL-1 in castrated male rats lacking vasopressinergic innervation of the lateral septum (Dantzer et al. 1991).

The mechanisms by which the brains vasopressinergic system is activated by cytokines and the way in which vasopressin interacts with the effect of cytokines on their target cells remain to be elucidated.

The potent activating effects of proinflammatory cytokines on pituitary-adrenal activity have already been mentioned. Glucocorticoids represent another class of key molecules in the regulation of sickness behavior. Adrenalectomy is accompanied by an increased sensitivity to the depressing effects of IL-1β and LPS on social exploration. This effect was mimicked by acute administration of the antiglucorticoid antagonist RU-38486 to intact mice and it was abrogated by the implantation of a corticosterone pellet in adrenalectomized mice (Goujon et al. 1995a).

Adrenalectomized animals implanted with a corticosterone pellet have constant levels of corticosterone but are unable to respond to administration of IL-1 by enhanced pituitary-adrenal activity. This phasic response to IL-1 appears to be important in regulating the behavioral effects of cytokines since the protection offered by a corticosterone pellet that ensured plasma levels of corticosterone intermediate between normal and stress levels was effective only against low but not high doses of IL-1β.

The effects of glucocorticoids on the behavioral activity of IL-1 and LPS are likely to be mediated by both peripheral and central mechanisms. In line with this prediction, adrenalectomized mice were found to be more sensitive to the depressing effects of a central injection of IL-1β on social exploration, and icv administration of the glucocorticoid receptor antagonist RU-38486 mimicked this effect (Goujon et al. 1995b). The mechanism underlying these effects is represented by the down regulation exerted by glucocorticoids on the synthesis and release of cytokines by both peripheral immune cells and brain cells and by the up regulation of the type II IL-1 receptor that acts as a decoy target (Goujon et al. 1996).

Pathophysiological Implications of the Behavioral Effects of Cytokines

The demonstration that the immune system is able to influence behavior and mental states has important implications for our understanding of the relationships between psychological factors and disease. In the case of cancer, for example, such psychological features as the feelings of hopelessness and helplessness that are commonly associated with the onset and progression of the disease might be secondary to the brain effects of factors released from immune or tumor cells during the early stage of the tumoral process. The same possibility applies to the relationship between psychological factors and autoimmune diseases. The possible causal role of cytokines in the mental and behavioral symptoms that occur in various pathological conditions has hardly been investigated, except in a few cases, such as infection and fever, cachexia, AIDS dementia complex, chronic fatigue syndrome and depression.

There is ample evidence demonstrating that proinflammatory cytokines are responsible for the development of subjective and behavioral symptoms of sickness during infection with a bacterial or viral pathogen. For instance, patients treated with IFN-α showed fever, anorexia, fatigue, headache, myalgia and arthralgia. These symptoms culminated in lethargy and withdrawal from the surroundings. The same symptoms were observed in volunteers injected with low doses of LPS.

The possibility that the release of cytokines accounts for more subtle changes in cognition and performance has been assessed by Smith et al. (1988). On the basis of earlier work showing that infection with upper respiratory viruses decreased the efficiency with which psychomotor tasks were performed, these researchers injected IFN-α in volunteers of both sexes. Volunteers injected with the larger dose were significantly slower at responding in a reaction time task when they were uncertain when the target stimulus would appear. Simultaneously, they displayed hyperthermia and experienced feelings of illness. However, they were not impaired on a pursuit tracking task or syntactic reasoning task. These effects were similar to the alterations in performance observed in patients with influenza. The possibility that proinflammatory cytokines have relatively specific effects on cognitive processes has been further investigated in animal models. IL-1β, but not IL-6, impaired the learning of a spatial navigation task in rats (Oitzl et al. 1993). However, the possible interference of sickness-induced altered performance with the task requirements was not systematically examined in this report. A similar deficit in spatial learning was observed in mice injected with IL-1β or infected with the pathogenic agent *Legionella pneumophilia* (Gibertini et al. 1995). Interference of cytokines with formation of new memories was also demonstrated in an autoshaping task in which rats learned to press a lever that was introduced into the cage before food delivery (Aubert et al. 1995b). These effects of cytokines were indepent of their pyrogenic activity since they were observed whether body temperature increased or decreased in response to the treatment under study.

There has been much speculation on the possible pathogenic role of cytokines in the chronic fatigue syndrome (CSF). Always feeling tired is a common complaint in patients afflicted with a viral infection and represents the core symptom of the so-called postviral fatigue syndrome. CSF patients feel the same but in the absence of any persistent viral infection (Kendell 1991). Their symptoms are real, pervasive and often incapacitating. The fact that a substantial proportion of these patients fulfill criteria for major depression and other psychiatric illness does not facilitate the classification of this disorder. Whatever the case, and in view of the similarities between the subjective effects of cytokines and the symptoms reported by CSF patients, many researchers have looked for possible hyperproduction of cytokines in this condition. Elevated plasma levels of IL-2 and IL-6 have been reported in one study of CSF patients (Lloyd et al. 1991), but this result has not been found in other studies. Such inconsistent results can be easily explained by technical problems associated with the detection of cytokines in biological fluids and the poor correlation between plasma levels of cytokines and local activity of these mediators. A better way of assessing peripheral cytokine function is to study the ability of peripheral blood mononuclear cells (PBMC) to produce cytokines when put in culture and stimulated with LPS. Using such a strategy, Chao et al. (1991) reported hyperproduction of IL-6 and TNF-α and a lower production of the inhibitory cytokine TGF-β in CSF patients. However, it is important to note that abnormalities, if any, of the cytokine network are not necessarily present at the periphery but might instead preferably affect cytokines that are expressed in the central nervous system (Dantzer 1994b, Volmer-Conna et al. 1998). Although this possibility is much more difficult to test, it has inspired quite elaborate *ad hoc* speculation (Goldstein 1993).

Besides their common occurrence in CSF patients, lack of energy and loss of interest are very frequent in depressed patients. These symptoms are actually incorporated in the basic description of depressive episodes. The tenth revision of the International Classification of Disease begins with the statement that "the subject suffers from a lowering of mood, reduction of energy, and decrease in activity. Capacity for enjoyment, interest and concentration are impaired, and marked tiredness after even minimum effort is common." The possibility that activation of peripheral blood monocytes and T lymphocytes plays a role in the pathophysiology of major depression has been proposed by Maes et al. (1995). In addition to the evidence pointing out the profound effects of cytokines on behavior and the hypothalamic-pituitary-adrenal axis, this hypothesis is based on the observation of an increased production of cytokines by monocytes and T-lymphocytes of depressed patients. For example, patients with major depression have elevated plasma levels of acute phase proteins and increased concentrations of IL-6 and its soluble receptor, with a close relationship between IL-6 levels and acute phase proteins. However, more research is still needed before a role of immune products in the pathogenesis of depressive symptoms can be accepted. The observed immune alterations might be a trait rather than a state marker of depression, since they persist even when depressive symptoms regress. In addition, the possible contribution of antidepressant treatment to the changes in immune functions observed in depressed patients remains to be established.

At the experimental level, the hypothesis of common mechanisms mediating the brain effects of cytokines and symptomatology of depression prompted a renewal of interest in the investigation of analogies between immune stressors and other stressors. The pattern of neurochemical changes that is observed in animals subjected to immunologic challenge is reminescent of that provoked by stressors (Dunn et al. 1989; Zalcman et al. 1991). In the same manner, the behavioral consequences of immunological activation share some similarity with those induced by stressors. This similarity is found even when the actions of cytokines in relatively specific models of animal depression are considered. For example, antigenic stimulation with sheep red blood cells decreased self stimulation from the nucleus accumbens but not from the substantia nigra in rats at the time of the peak immune response, and this effect was similar to what is observed following exposure to inescapable electric shock (Zacharko et al. 1997). In the same manner, administration of LPS reduced saccharin preference in rats (Yirmiya 1996). These similarities encouraged researchers to test the effects of antidepressant treatment on the behavioral effects of cytokines and, reciprocally, the effects of treatments blocking cytokine actions on animal models of depression. Chronic but not acute treatment with imipramine abrogated the LPS-induced decrease in saccharin preference (Yirmiya 1996), whereas intracerebroventricular administration of IL-1ra blocked the impairing effects of acute exposure to a stressor on the ability to learn an escape response (Maier and Watkins 1995). Although there is still a long way to go, these findings are strongly supportive of the hypothesis of common mechanisms between cytokine-induced sickness and some forms of depression.

Cachexia and anorexia are proeminent features of chronic infection and cancer. PBMC isolated from cancer patients release more TNF-α in response to stimulation than cells from healthy controls. Elevated serum levels of TNF-α have also been reported in patients with a variety of malignancies. The possible involvement of TNF-α in the metabolic changes occurring in malnutritional status has been deduced from the demonstration of an enhanced release of this cytokine by stimulated PBMC from short-term, acutely starved subjects and non-stimulated PBMC from patients afflicted with anorexia nervosa. Anorexia, weight loss and wasting are important causes of morbidity in AIDS patients. The pathophysiology of the AIDS wasting syndrome includes three components: impaired nutrient intake, decreased nutrient absorption and metabolic alterations (Grunfeld and Kotler 1992). The possibility that cytokines play a key role in the AIDS wasting syndrome has been proposed on the basis of two different sets of findings: 1) HIV infection is accompanied by activation of several cytokine genes and 2) many characteristic features of the AIDS wasting syndrome are similar to those seen in other chronic infections and cancer and are due to cytokines. Infection and activation of $CD4^+$ lymphocytes and macrophages by HIV are asociated with the production of a range of cytokines, especially TNF-α. Elevated levels of this cytokine are found in the plasma of HIV-infected patients. In addition, PBMC of HIV-infected individuals produce higher levels of TNF-α. Other cytokines, including IL-1, IL-6 and TGF-β are also hyperexpressed in HIV-1 infection.

HIV-1 is able to enter the CNS and infect macrophages and microglial cells (Vintners and Anders 1990). However, the small number of infected cells contrasts with the extensive amount of functional alterations, of which the most well known is AIDS-related dementia. Many AIDS patients present a primary neuropsychiatric disorder, including slowness in thinking, forgetfulness, cognitive changes, drowsiness, weakness and difficulties in concentrating. These neurological abnormalities occur in as many as 80% of all AIDS patients, cannot generally be attributed to opportunistic infections and may even precede most other clinical signs of infection with HIV. Since HIV-1 is well known to induce the synthesis of TNF-α, IL-1, IL-6 and IL-2 from microglia, it has been proposed that these cytokines contribute to the neurological symptoms observed in AIDS patients. Examination of brains from HIV-1-infected patients by immunocytochemistry revealed higher expression of IL-1 in endothelial cells and TNF-α in macrophages and microglial cells. TGF-β has also been identified in HIV-1-infected brain and appears to be localized exclusively to tissue areas with pathological abnormalities (Vitkovic 1997).

Conclusion

Sufficient evidence is now available to accept the concept that cytokines are interpreted by the brain as internal signals of sickness. Sickness can actually be considered as a motivation, that is, a central state that organizes perception and action in face of the threat of infection by pathogen microorganisms. A sick individual does not have the same priorities as a well one, and this reorganization of priorities is mediated by the effects of cytokines on a number of peripheral and central targets. The elucidation of the mechanisms that are involved in these effects should provide new insight into the way sickness and recovery processes are represented in the brain.

References

Aubert A, Goodall G, Dantzer R (1995a) Compared effects of cold ambient temperature and cytokines on macronutrient intake in rats. Physiol Behav 57:869–873

Aubert A, Vega C, Dantzer R, Goodall G (1995b) Pyrogens specifically disrupt the acquisition of a task involving cognitive processing in the rat. Brain Behav Immunity 11:107–118

Aubert A, Goodall G, Dantzer R, Gheusi G (1997a) Differential effects of lipopolysaccharide on pup retrieving and nest building in lactating mice. Brain Behav Immunity 11:107–118

Aubert A, Kelley KW, Dantzer R (1997b) Differential effects of lipopolysaccharide on food hoarding behavior and food consumption in rats. Brain Behav Immunity 11:229–238

Baumann H, Gauldie J (1994) The acute phase response. Immunol Today 15:74–80

Bluthé RM, Dantzer R, Kelley KW (1992) Effects of interleukin-1 receptor antagonist on the behavioral effects of lipopolysaccharide in rat. Brain Res 573:318–320

Bluthé RM, Pawlowski M, Suarez S, Parnet P, Pittman Q, Kelley KW, Dantzer R (1994a) Synergy between tumor necrosis factor α and interleukin-1 in the induction of sickness behaviour in mice. Psychoneuroendocrinology 19:197–207

Bluthé RM, Walter V, Parnet P, Layé S, Lestage J, Verrier D, Poole S, Stenning BE, Kelley KW, Dantzer R (1994b) Lipopolysaccharide induces sickness behavior in rats by a vagal mediated mechanism. C R Acad Sci Paris/Sciences de la Vie 317:499–503

Bluthé RM, Michaud B, Kelley KW, Dantzer R (1996a) Vagotomy atenuates behavioural effects of interleukin-1 injected peripherally but not centrally. Neuroreport 7:1485–1488

Bluthé RM, Michaud B, Kelley KW, Dantzer R (1996b) Vagotomy blocks behavioural effects of interleukin-1 injected via the intraperitoneal route but not via other systemic routes. Neuroreport 7:2823–2827

Bret-Dibat JL, Bluthé RM, Kent S, Kelley KW, Dantzer R (1995) Lipopolysaccharide and interleukin-1 depress food-motivated behavior in mice by a vagal-mediated mechanism. Brain Behav Immunity 9:242–246

Buttini M, Boddeke H (1995) Peripheral lipopolysaccharide stimulation induces interleukin-1β messenger RNA in rat brain microglial cells. Neuroscience 65:523–530

Chao CC, Janoff EN, Hu S (1991) Altered cytokine release in peripheral blood mononuclear cell cultures from patients with the chronic fatigue syndrome. Cytokine 3:292–295

Cunningham E Jr, Wada E, Carter DB, Tracey DE, Battey JF, De Souza EB (1992) In situ histochemical localization of type I interleukin-1 receptor messenger RNA in the central nervous system, pituitary, and adrenal gland of the mouse. J Neurosci 12:1101–1114

Dantzer R (1994a) How do cytokines say hello to the brain? Neural versus humoral mediation. Eur Cytokine Network 5:271–273

Dantzer R (1994b) Current studies on the neurobiology of chronic fatigue syndrome. Encephale 20, Suppl. 3:597–602

Dantzer R, Bluthé RM, Kelley KW (1991) Androgen-dependent vasopressinergic neurotransmission attenuates interleukin-1 induced sickness behavior. Brain Res 557:115–120

Dantzer R, Bluthé RM, Kent S, Goodall G (1993) Behavioral effects of cytokines: An insight into mechanisms of sickness behavior. In: De Souza E (ed) Methods in neuroscience. Academic Press, Orlando, pp 130–149

Dinarello CA (1996) Biologic basis for interleukin-1 in disease. Blood 87:2095–2147

Dunn AJ, Powell ML, Meitin C, Small PA (1989) Virus infection as a stressor: influenza virus elevates plasma concentrations of corticosterone, and brain concentrations of MHPG and tryptophan. Physiol Behav 45:591–594

Fleshner M, Goehler LE, Hermann J, Relton JK, Maier SF, Watkins LR (1995) Interleukin-1β induced corticosterone elevation and hypothalamic NE depletion is vagally mediated. Brain Res Bull 37:605–610

French RA, Zachary JF, Dantzer R, Frawley LS, Chizzonite R, Parnet P, Kelley KW (1996) Dual expression of p80 type I and p68 type II interleukin-1 receptors on anterior pituitary cells synthesizing growth hormone. Endocrinology 137:4027–4036

French RA, Chizzonite R, Dantzer R, Parnet P, Bluthé RM, VanHoy RW, Zachary JF, Kelley KW (1999) Expression and localization of p80 and p68 interleukin-1 receptor proteins in the brain of adult mice. J Neuroimmunol 93:194–202

Gatti S, Bartfai T (1993) Induction of tumor necrosis factor-α mRNA in the brain after peripheral endotoxin treatment; comparison with interleukin-1 family and interleukin 6. Brain Res 624: 291–294

Gibertini M, Newton C, Friedman H, Klein TW (1995) Spatial learning impairment in mice infected with *Legionella pneumophila* or administered exogenous interleukin-1β. Brain Behav Immunity 9:113–128

Goldstein JA (1993) Chronic fatigue syndrome: the limbic hypothesis. Harworth Medical Press, Binghamton

Goujon E, Parnet P, Aubert A, Goodall G, Dantzer R (1995 a) Corticosterone regulates behavioral effects of lipopolysaccharide and interleukin-1β in mice. Am J Physiol 269:R154–R159

Goujon E, Parnet P, Cremona S, Dantzer R (1995 b) Endogenous glucocorticoids down regulate central effects of interleukin-1β on body temperature and behavior in mice. Brain Res 702:173–1280

Goujon E, Parnet P, Layé S, Combe C, Dantzer R (1996) Adrenalectomy enhances proinflammatory cytokines gene expression in the spleen, pituitary and brain of mice in response to lipopolysaccharide. Mol Brain Res 36:53–62

Grunfeld C, Kotler DP (1992) Pathophysiology of the AIDS wasting syndrome. AIDS Clin Rev 191–224

Haour F, Ban E, Baran D, Millon G, Fillion G (1990) Brain interleukin-1 receptors in the mouse: characterization and modulation after lipopolysaccharide (LPS) injection. Progr Neuroimmunoendocrinol 3:196–204

Hart BL (1988) Biological basis of the behavior of sick animals. Neurosci Biobehav Rev 12:123–137

Hopkins SJ, Rothwell NJ (1995) Cytokines and the nervous system, I: expression and recognition. Trends Neurosci 18:83–87

Katsuura K, Arimura A, Koves K, Gottschall PE (1990) Involvement of organum vasculosum of laminae terminalis and preoptic area in interleukin-1β induced ACTH release. Am J Physiol 258:E163–E171

Kendell RE (1991) Chronic fatigue, viruses and depression. Lancet 337:160–162

Kent S, Bluthé RM, Dantzer R, Hardwick AJ, Kelley KW, Rothwell NJ, Vannice JL (1992 a) Different receptor mechanisms mediate the pyrogenic and behavioral effects of interleukin-1. Proc Natl Acad Sci USA 89:9117–9120

Kent S, Bluthé RM, Kelley KW, Dantzer R (1992 b) Sickness behavior as a new target for drug development. Trends Pharmacol Sci 13:24–28

Kent S, Bret-Dibat JL, Kelley KW, Dantzer R (1995) Mechanisms of sickness-induced decreases in food-motivated behavior. Neurosci Biobehav Rev 20:171–175

Kluger MJ (1979) Fever: its biology, evolution and function. Princeton University Press, Princeton.

Kluger MJ (1991) Fever: role of pyrogens and cryogens. Physiol Rev 71:93–127

Konsman JP, Kelley KW, Dantzer R (1999) Temporal and spatial relationships between lipopolysaccharide-induced expression of Fos, interleukin-1β and inducible NO synthase in rat brain. Neuroscience 89:535–548

Layé S, Parnet P, Goujon E, Dantzer R (1994) Peripheral administration of lipopolysaccharide induces the expression of cytokine transcripts in the brain and pituitary of mice. Mol Brain Res 27:157–162

Layé S, Bluthé RM, Kent S, Combe C, Médina C, Parnet P, Kelley KW, Dantzer R (1995) Subdiaphragmatic vagotomy blocks the induction of interleukin-1β mRNA in the brain of mice in response to peripherally administered lipopolysaccharide. Am J Physiol 268:R1327–R1331

Licinio J, Wong ML, Gold PW (1991) Localization of interleukin-1 receptor antagonist mRNA in rat brain. Endocrinology 129:562–564

Lloy A, Hickie I, Brockman A, Dwyer J, Wakefield D (1991) Cytokine levels in serum and cerebrospinal fluid in patients with chronic fatigue syndrome and control subjects. J Infect Dis 164: 1023–1024

Maes M, Smith R, Scharpe S (1995) The monocyte-T-lymphocyte hypothesis of major depression, Psychoneuroendocrinology 20:111–116

Maier SE, Watkins LR (1995) Intracerebroventricular interleukin-1 receptor antagonist blocks the enhancement of fear conditioning and interference with escape produced by unescapable shock. Brain Res 695:279–282

Miller NE (1964) Some psychophysiological studies of motivation and of the behavioral effects of illness. Bull Brit Psychol Soc 17:1–20

Oitzl MS, van Oers H, Schöbitz B, De Kloet ER (1993) Interleukin-1β, but not interleukin-6, impairs spatial navigation learning. Brain Res. 613:160–163

Parnet P, Brunke DL, Goujon E, Demotes-Mainard J, Biragyn A, Arkins S, Dantzer R, Kelley KW (1993) Molecular identification of two types of interleukin-1 receptors in the murine pituitary gland. J Neuroendocrinol 5:213–219

Parnet P, Amindari S, Wu C, Brunke-Reese DL, Goujon E, Weyhenmeyer JA, Dantzer R, Kelley KW (1994) Expression of type I and type II interleukin-1 receptors in the mouse brain. Mol Brain Res 27:63–70

Plata-Salaman CR (1996) Anorexia during acute and chronic disease. Nutrition 12:69–78

Smith A, Tyrrell D, Coyle K, Higgins P (1988) Effects of interferon alpha in man: a preliminary report. Psychopharmacology 96:414–416

Stitt JT (1985) Evidence for the involvement of organum vasculosum laminae terminalis in the febrile response of rabbits and rats. J Physiol London 368:501–511

Takao T, Tracey DE, Mitchell WM, De Souza EB (1990) Interleukin-1 receptors in mouse brain: characterization and neuronal localization. Endocrinology 127:3070–3078

Van Dam AM, Brouns M, Louisse S, Berkenbosch F (1992) Appearance of interleukin-1 in macrophages and ramified microglia in the brain of endotoxin-treated rats: a pathway for the induction of non specific symptoms of sickness. Brain Res 588:291–296

Vintners HV, Anders KH (1990) Neuropathology of AIDS. CRC Press, Boca Raton, Florida.

Vitkovic L (1997) Neuropathogenesis of HIV-1 infection: interactions between interleukin-1 and transforming growth factor-β1. Mol Psychiat 2:111–112

Vollmer-Conna U, Lloyd A, Hickie I, Wakefield D (1998) Chronic fatigue syndrome: an immunological perspective. Aust N Z J Psychiatry 32:523–527

Wan W, Wetmore L, Sorensen CM, Greenberg AH, Nance DM (1994) Neural and biochemical mediators of endotoxin and stress-induced c-fos expression in the rat brain. Brain Res Bull 34:7–14

Watkins LR, Wiertelak E., Goehler LE, Mooney-Heiberger K, Martinez J, Furness L, Smith KP, Maier SF (1994a) Neurocircuitry of illness-induced hyperalgesia. Brain Res 639:283–299

Watkins LR, Wiertelak EP, Goehler LE, Smith KP, Martin D, Maier SF (1994b) Characterization of cytokine-induced hyperalgesia. Brain Res 654:15–26

Watkins LR, Goehler LE, Relton J, Tartaglia N, Silbert L, Martin D, Maier SF (1995) Blockade of interleukin-1 induced hyperthermia by subdiaphragmatic vagotomy: evidence for vagal mediation of immune-brain communication. Neurosci Lett 183:27–31

Yirmiya R (1996) Endotoxin produces a depressive-like episode in rats. Brain Res 711:163–174

Zacharko RM, Zalcman S, Macneil G, Andrews M, Mendalla PD, Anisman H (1997) Differential effects of immunologic challenge on self-stimulation from the nucleus accumbens and the substantia nigra. Pharmacol Biochem Behav 58:881–886

Zalcman S, Shanks N, Anisman H (1991) Time-dependent variations of central norepinephrine and dopamine following antigen administration. Brain Res 557:69–76

Central Origin of IL-1β Produced During Peripheral Inflammation: Role of Meninges

B. Vernet-der Garabedian, Y. Lemaigre-Dubreuil, N. Delhaye-Bouchaud, and J. Mariani

Summary

Modulation of IL-1β in the central nervous system is well documented. In the murine model of systemic inflammation induced by peripheral injection of lipopolysaccharide (LPS), data consensually demonstrated an increase of brain IL-1β mRNA after LPS treatment. Expression at the protein level has been much more widely debated, mainly in the cases of demonstrations assessed by methods using global extracts of brain tissues. We hypothesized that these discrepancies may originate from contaminations by meningial tissue. Indeed, meninges belong to the periphery and are highly responsive to LPS. We have evaluated the contribution of this tissue in the IL-1β response of hippocampus, cortex and cerebellum from mice treated with LPS or RPMI. Comparison of IL-1β mRNA levels in extracts from structures collected with meninges to structures cautiously stripped from meningial membranes, vessels and choroid plexus showed that two thirds of the signal had a meningial origin. The contribution of the brain parenchyma itself was likely overestimated, as meninges remained in the brain foldings despite a careful dissection. Immunohistochemical studies of IL-1β on brain sections corroborated these results: staining was only observed in a few large cells located in the membranes surrounding the whole brain. No obvious staining was found in the parenchyma, at least with the doses of LPS we used.

These results show that the meningial tissue can constitute the major source of contamination of nervous parenchyma IL-1β by peripheral IL-1β if meninges are not removed. They tempt to the greatest care when analyzing data obtained from global extracts of the brain.

Introduction

Interleukin-1β (IL-1β) is one of the key components of the inflammatory reaction developed to constrain viral or bacterial infections and to repair lesioned tissues (Dinarello 1994). IL-1β is responsible for many of the effects associated with peripheral inflammation such as fever, sleep, behavioral changes, and activation and inhibition of the hypothalamic-pituitary-adrenal axis and of hypothalamic-pituitary-gonadal axis, respectively. It is now clear

Patterson/Kordon/Christen (Eds.)
Neuro-Immune Interactions in Neurologic and Psychiatric Disorders

that all of these peripheral manifestations are centrally controlled and that brain IL-1β participates to the cross-talk between the peripheral compartment and the central nervous system. Brain expression of IL-1β mRNA and its modulation by an acute systemic inflammation has been extensively documented in different animal models (Hopkins and Rothwell 1995; Rothwell and Hopkins 1995). Increased brain level of IL-1β mRNA has been demonstrated after lipopolysaccharide (LPS) treatment in rats or mice. At the protein level, localization within brain remains somewhat controversial (Buttini and Boddeke 1995; Nakamori et al. 1994; Quan et al. 1994; van Dam et al. 1992; Wong et al. 1996) and data from techniques using global brain extracts are conflicting (Goujon et al. 1996; Takao et al. 1993; Vernet-der Garabedian et al. 1998). Techniques such as RT-PCR, RIA, and ELISA, based on global extract analysis, are widely used to demonstrate the presence of a molecule in a tissue. In the case of brain structures, they may lead to erroneous interpretations depending on how the tissue samples were prepared. Problems may originate from contaminations due to the meninges or plexus choroid, both of which belong to the peripheral compartment. During brain dissection, the most external layer of meninges, the dura mater, is taken off with the skull while the remaining meninges, i.e., the pia mater and the arachnoid, wrap the brain with a thin membrane and penetrate deep into the fissures. Without care during preparation, nervous tissue, i.e., nervous parenchyma samples, could be contaminated by meningial tissue, which, moreover, is highly responsive to LPS (Perry et al. 1995).

The aim of the present study was to evaluate the contribution of the meningial tissue to the IL-1β response of brain after a peripheral LPS stimulation. Using RT-PCR, we determined IL-1β mRNA levels of total RNA extracts from different brain regions (hippocampus, cerebral cortex or cerebellum) collected with meninges or plexus choroid and after careful dissection to remove them, before and after LPS treatment. We have also localized the produced IL-1β by immunocytochemistry in the same structures after LPS treatment.

Methods

Thirty-day-old C57BL/6 mice (Janvier Laboratory) were injected i.p. with purified LPS from *E. coli*, type 0127-B8 (Sigma) in RPMI as vehicle. They received one injection of LPS (0.5 μg/g body weight) or RPMI and were sacrificed by cervical dislocation two hours later. We chose this duration of treatment on the basis of our previous work showing that the increase of brain IL-1β mRNA peaked two hours after LPS treatment (Lemaigre-Dubreuil et al. 1996). Each experiment was repeated three to five times.

LPS and vehicle-treated mice were divided in two groups of five animals, respectively, and rapidly dissected. For one group, brain structures (cerebellum, hippocampus and cerebral cortex) were collected on ice in a RPMI bath and immediately frozen in liquid nitrogen. For the second group of animals, the same brain structures were carefully stripped (under binocular microscope) from the visible meninges (membranes, vessels and choroid plexuses) before freezing. Meninges were kept and pooled. Tissues were stored at −80 °C until used. Total RNA was isolated with RNA^B (Bioprobe). Concentration and qualitv were checked by spectrophotometry and electrophoresis on agarose gels. RNA was reverse transcribed by the Riboclone® cDNA synthesis system with Oligo $(dT)_{15}$ primers and AMV reverse transcriptase (Promega). PCR amplification was carried out in the presence of control G3PDH primers or mouse IL-1β amplimer sets and competitor DNA for IL-1β (Clontech). The PCR reaction contained in a final volume of 50 µL, 10 mM Tris-HCI, 50 mM KCl, 1.5 mM $MgCl_2$, 0.01% Gelatin, dNTP 0.2 nM each, primers 0.4 µM each, 1 µg cDNA and 2.0 U Taq DNA Polymerase. Amplification was carried out on a Perkin Elmer DNA Thermal Cycler. The PCR reactions were carried out under non-saturated conditions and control reactions were performed in the absence of reverse transcriptase to rule out genomic DNA contamination. The PCR products were electrophoresed on 2% agarose gels, and the ethidium bromide stained bands were analyzed. The resulting IL-1β data were normalized using the corresponding competitor DNA and G3PDH values.

Frozen air-dried sections of brain were fixed with 75% acetic acid plus 25% ethanol for 12 min, progressively hydrated in 70, 50 and 30% ethanol and washed in PBS. Endogenous peroxidases were inhibited with 0.3% H_2O_2, 10% methanol in PBS for 30 min and the sections were incubated with PBS supplemented with 0.2% Triton X-100 and 10% goat serum for 20 min to block nonspecific binding sites. Sections were then incubated with 5 µg/ml of IL-1β specific antibody (hamster anti-mouse IL-1β monoclonal antibody, Genzyme) diluted in PBS plus 0.1% bovine serum albumin overnight at 4 °C. After washing, biotinylated goat anti-hamster antibody IgG, affinity purified (Vector) were applied at a 1/400 dilution for three hours. Staining was visualised using the Vectastain ABC kit with diaminobenzidine (Sigma) as chromogen. Control sections were exposed to PBS plus 0.1% bovine serum albumin alone. To get a positive control, anti-IL-1β antibodies were tested in the same conditions on cultured macrophages on which positive staining was observed after LPS stimulation (data not shown).

Results

Basal Levels of IL-1β mRNA Expression Within the Brain

Basal levels of IL-1β mRNA expression (as judged from the amplified IL-1β cDNA) were almost the same in all the studied brain structures of C57BL/6 mice, i.e., hippocampus, cerebral cortex and cerebellum. A careful stripping of the structures from visible meningial membranes and choroid plexuses had no significant effect on the basal values (Fig. 1).

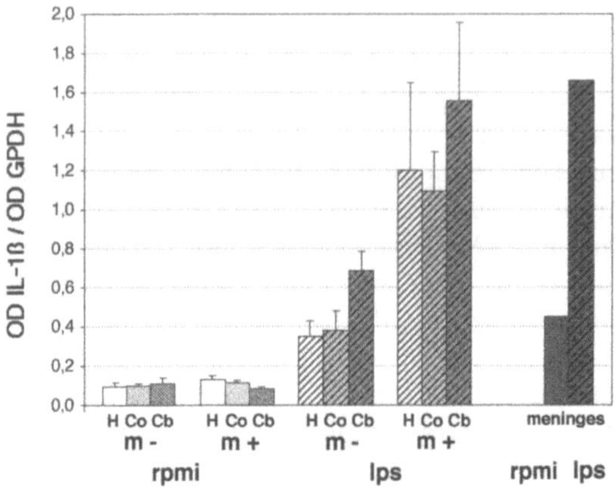

Fig. 1. Amplification of IL-1β mRNA levels by the presence of meningial tissue in brain structures after peripheral LPS treatment. Hippocampus (white bars), cortex (light gray bars), cerebellum (medium gray bars) and isolated meninges (dark gray bars) of C57BL/6J mice were collected with (m+) or without (m−) meninges after RPMI (unstriped bars) and LPS treatments (striped bars). IL-1β mRNA levels were quantified as described in the Methods section. Values are means of three to five samples ± SEM. Meninges were pooled tissues

IL-1β mRNA Expression After Peripheral LPS Treatment in Brain Structures and Isolated Meninges

We measured IL-1β mRNA expression in stripped (without meninges) and unstripped (with meninges) brain structures and in isolated meninges from mice sacrificed two hours after i.p. LPS injection. As shown in Figure 1, LPS treatment compared to RPMI increased IL-1β mRNA from 3.6 to 6.1-fold in tissues without meninges, from 9.2 to 17.8-fold in tissues with meninges and 3.7-fold in the isolated meninges. Cerebellum appeared to be the most responsive structure. Comparison of unstripped to stripped structures revealed that the presence of meninges amplified IL-1β mRNA response to LPS 3.4-, 2.9- and 2.3-fold (mean: 2.9), respectively, in hippocampus, cortex and cerebellum.

Immunohistochemical Studies of IL-1β Expression in Brain

As shown in Figure 2, a small number of large cells looking like macrophages were found immunopositive in the pia mater surrounding all the brain. By contrast, no obvious staining was observed elsewhere in the parenchyma after LPS stimulation. No staining was observed in the untreated animals.

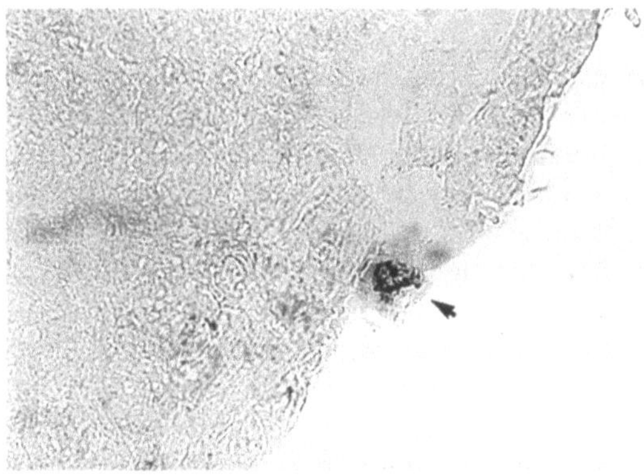

Fig. 2. Localization of IL-1β expression in the brain. Immunohistochemical staining of a sagittal brain section of LPS treated mice. A cell looking like a macrophage (arrow) and located in the pia mater surrounding the brain shows positive reactivity for the monoclonal anti-mouse IL-1β antibody. No staining is apparent in the nervous parenchyma (magnification X 900)

Discussion

These results clearly show that meninges could be responsible for about two-thirds of the IL-1β signal of the brain response to a peripheral inflammatory stimulus. That means that the contribution of brain parenchyma itself to IL-1β mRNA signal could be largely overestimated if meninges are not removed during extract preparation.

From the data presented here, the contribution of the parenchyma ranges around one-third and this value could be still overestimated. Indeed, it is very difficult to properly strip out all the meninges from the cerebral structures, particularly in the case of the cerebellum, which is a highly folded structure. Beside meninges, a non-negligible source of contamination is the brain vasculature. First, vessels contain circulating immune cells from the periphery. Second, endothelial cells have been demonstrated to respond to inflammatory stimuli such as LPS in producing IL-1β (Fabry et al. 1993), and the brain is one of the best irrigated tissues. It is possible to minimize the contaminations due to meninges or choroid plexus by cautious stripping and those due to circulating cells by perfusion of the brain with saline solution. However, the endothelium lining the parenchymatous vasculature cannot be removed.

We have performed immunohistochemical studies on brain sections using a monoclonal anti-IL-1β antibody. Beside the few immunostained macrophages localized in the pia mater of some brain sections, no obvious staining was observed in the parenchyma at the dose of LPS we used. These results agree with previous studies performed on the *staggerer* cerebellar mutant mice. In this mutant a prominent overexpression of IL-1β mRNA (Lemaigre-Dubreuil et al. 1996) and an increased level of IL-1β (Vernet-der Garabedian et al. 1998) were demonstrated in the cerebellar tissue. However, immunostained macrophages were also found localized in the peripheral meninges and choroid plexus (Saisset et al., unpublished data).

The present results question the presence of IL-1β in the brain after systemic inflammation.

Previous studies from other laboratories demonstrate modulation of IL-1β expression in the brain of mice or rat after peripheral LPS treatment (Buttini and Boddeke 1995; Goujon et al. 1996; Nguyen et al. 1998; Quan et al. 1994; van Dam et al. 1992). Discrepancies between our data and those from others may be due to differences in stimulation by LPS (origin, doses). High doses (2.5 to 5 mg/kg body weight) have been reported to induce IL-1β changes throughout the brain at the mRNA or protein levels, whereas lower doses only affected meninges and brain regions with fenestrated endothelial capillaries such as the circumventricular organs and choroid plexus (Buttini and Boddeke 1995; van Dam et al. 1992). LPS is a complex glycolipid that pleiotropically stimulates the cells implicated in the inflammatory response. Injected at high doses it could modify the blood brain-barrier permeability, cross the cerebral endothelium and directly stimulate the resident brain monocyte lineage cells (Abbott and Revest 1991; Lynn and Golenbock 1992). We were interested in the reaction in the brain to a peripheral inflammation; thus LPS was used at a low dose (0.5 mg/kg) that was nevertheless sufficient to stimulate splenic macrophages (16.7-fold above basal level; data not shown) and to induce behavioral changes in mice.

In conclusion, our results show that the greatest care must be taken when analyzing data obtained from crude or global extracts of the brain. Such methods (RT-PCR, ELISA, RIA) are certainly not sufficient to demonstrate the presence of a molecule in the nervous parenchyma when this molecule is also produced by the meninges. They should be associated with cytological techniques, such as immunohistochemistry and in situ hybridization, to identify and localize the cells producing the studied molecule.

Acknowledgments. We are grateful to Dr. Paul Derer for his help in the brain section interpretation. We wish to thank Paulette Bouquet for her help in performing histological technique and Colette Chianale for the breeding of the mice. This work was supported by a grant from the European Community (grant BIOMED. Cybrainet n° PL 962492).

References

Abbott NJ, Revest PA (1991) Control of brain endothelial permeability. Cerebrovasc Brain Metab Rev 3:39–72

Buttini M, Boddeke H (1995) Peripheral lipopolysaccharide stimulation induces interleukin-1 beta messenger RNA in rat brain microglial cells. Neuroscience 65:523–530

Dinarello CA (1994) The biological properties of interleukin-1. Eur Cytokine Netw 5:517–531

Fabry Z, Fitzsimmons KM, Herlein JA, Moninger TO, Dobbs MB, Hart MN (1993) Production of the cytokines interleukin 1 and 6 by murine brain microvessel endothelium and smooth muscle pericytes. J Neuroimmunol 47(1):23–34

Goujon E, Parnet P, Layé S, Combe C, Dantzer R (1996) Adrenalectomy enhances pro-inflammatory cytokines gene expression, in the spleen, pituitary and brain of mice response to lipopolysaccharide. Mol Brain Res 36:53–62

Hopkins SJ, Rothwell NJ (1995) Cytokines and the nervous system 1: Expression and recognition. Trends Neurosci 18:83–88

Lemaigre-Dubreuil Y, Brugg B, Chianale C, Delhaye-Bouchaud N, Mariani J (1996) Over-expression of interleukin-1beta-converting enzyme mRNA in staggerer cerebellum. Neuroreport 7: 1777–1780

Lynn WA, Golenbock DT (1992) Lipopolysaccharide antagonists. Immunol Today 13:271–276

Nakamori T, Morimoto A, Yamaguchi K. Watanabe T, Murakami N (1994) Interleukin-1 beta production in the rabbit brain during endotoxin-induced fever. J Physiol 476.1:177–186

Nguyen KT, Deak T, Owens SM, Kohno T, Fleshner M, Watkins LR, Maier SF (1998) Exposure to acute stress induces brain interleukin- 1β protein in the rat. J Neurosci 18(6):2239–2246

Perry VH, Bell MD, Brown HC, Matyszak MK (1995) Inflammation in the nervous system. Curr Opin Neurobiol 5:636–641

Quan N, Sundar S, Weiss J (1994) Induction of interleukin-1 in various brain regions after peripheral and central injections of lipopolysaccharide. J Neuroimmunol 49:125–134

Rothwell NJ, Hopkins SJ (1995) Cytokines and the nervous system II: actions and mechanisms of action. Trends Neurosci 18:130–136

Takao T, Culp S, De Souza E (1993) Reciprocal modulation of interleukin-1beta (IL-1β) and IL-1 receptors by lipopolysaccharide (endotoxin) treatment in the mouse brain-endocrine-immune axis. Endocrinology 132:1497–1504

van Dam A-M, Brouns M, Louisse S, Berkenbosch F (1992) Appearance of interleukin-1 in macrophages and in ramified microglia in the brain of endotoxin-treated rats: a pathway for the induction of non-specific symptoms of sickness? Brain Res 588:291–296

Vernet-der Garabedian B, Lemaigre-Dubreuil Y, Delhaye-Bouchaud N, Mariani J (1998) Abnormal IL-1β cytokine expression in the cerebellum of the ataxic mutant mice *staggerer* and *Lurcher*. Mol Brain Res 62(2):224–227

Wong M-L, Bongiorno PB, Al-Shekhlee A, Esposito A, Khatri P, Licinio J (1996) IL-1beta, IL-1 receptor type I and iNOS gene expression in rat brain vasculature and perivascular areas. Neuroreport 7:2445–2448

Vitamin D, a Hormone Involved in the Control of Neuro-Immune Interactions in the Brain

P. Brachet, E. Garcion, C.N. Montero-Menei, S. Nataf, L. Sindji, I. Neveu, P. Naveilhan, D. Wion, and F. Darcy

1,25-Dihydroxyvitamin D3, a Pluripotent Hormone

Vitamin D is a secosteroid that can be formed by the cleavage of 7-dehydrocholesterol by UV light. The biologically active metabolite of vitamin D is generated following two hydroxylation steps, in position 25 in the liver and position 1α in the kidney. 1α,25-dihydroxyvitamin D3 (1,25-D3) binds to an intracellular receptor belonging to the nuclear hormone-receptor superfamily, which includes receptors for classical steroids, tri-iodothyronine (T3) or retinoic acid (RA). The vitamin D receptor (VDR) can form active heterodimers, for instance with the RXR receptor, which binds 9-cis RA. Binding of VDR to responsive elements present in regulatory regions of target genes results in transcriptional effects (for review, see Christakos et al. 1996; Haussler et al. 1997). In addition, 1,25-D3 exerts membrane-mediated effects such as the induction of rapid, transient changes of intracellular free calcium, which are observed in numerous cell types. 1,25-D3 is mainly known for its calcemic and phosphatemic action in intestine or bone. However, it is also active in the immune system, in which monocytes or activated lymphocytes express the VDR. 1,25-D3 promotes the differentiation of monocytes into macrophages and increases their phagocytic potential (Provvedini et al. 1986; Rook et al. 1986; Kreutz and Andreseen 1990). Conversely, the hormone inhibits the production of TNF-α and IL- 12 (Zarrabeitia et al. 1992; Lemire et al. 1994), an interleukin that favours the differentiation of CD4-positive lymphocytes towards the Th1 subtype. Likewise, the hormone down-regulates various cell surface proteins, such as the major histocompatibility complex (MHC) class II antigen, CD4, or the B7-2 costimulatory molecule (Rigby et al. 1990; Spittler et al. 1997; Clavreul et al. 1998). 1,25-D3 exerts an anti-proliferative action on T or B lymphocytes and inhibits more specifically the production of IFN-γ and IL-2, which are released by the Th1 lymphocytes (Reichel et al. 1987; Rigby et al. 1987). Hence, the hormone exerts various immunosuppressive effects that are evidenced in vivo by the fact that *prophylactic* 1,25–D3 treatments, performed at pharmacological doses, prevent the outcome of autoimmune diseases such as insulinitis or lupus erythematosus in genetically susceptible mice (Lemire et al. 1992; Mathieu et al. 1992). Likewise, 1,25-D3, when applied *during the immunization phase,* protects susceptible rodents against experimental allergic encephalomyelitis (Lemire and Archer 1990).

Patterson/Kordon/Christen (Eds.)
Neuro-Immune Interactions in Neurologic and Psychiatric Disorders

The Central Nervous System is a Target of 1,25-D3

Until recently, the possibility that 1,25-D3 might control neuroimmune or neuroendocrine interactions in the central nervous system (CNS) remained poorly investigated. The presence of VDR in the CNS was first suggested by the detection of 1,25-D3 high affinity binding sites in neurons, but also in non-neuronal structures such as the choroid plexus (Stumpf and O'Brien 1987; Bidmon et al. 1991). This finding was further confirmed by the identification of VDR-like immunoreactivity in the rodent brain (Clemens et al. 1988) and of VDR mRNA in the human hippocampus (Sutherland et al. 1992). The transit of 1,25–D3 across the blood-brain barrier is possible, although limited, and vitamin D metabolites are found in the cerebrospinal fluid (Gascon-Barré and Huet 1983; Balabanova et al. 1984; Pardridge et al. 1985). However, activated microglial cells, like peripheral macrophages (Reichel et al. 1987), express a 1α-hydroxylase activity and can produce 1,25-D3 from its precursor, at least in vitro (Neveu et al. 1994 a). This raises the possibility that a similar hydroxylation step can take place in vivo, for instance in response to pro-inflammatory stimuli, implying that 1,25-D3 can be synthesized in the brain parenchyma.

Neuroprotective Actions of 1,25-D3

In vivo, 1,25-D3 up-regulates calbindin D28k or parvalbumin in motoneurons (Alexianu et al. 1998). We have shown that it also increases the specific activity of γ-glutamyl transpeptidase (Garcion et al. 1996), a key enzyme involved in the production of glutathione. Therefore, 1,25-D3 appears to influence the calcium-buffering potential of some neurons and the anti-oxidant defenses of astrocytes. The hormone also regulates the expression of the low affinity NGF receptor, LNGFR or p75 (Naveilhan et al. 1996 a). This receptor is structurally related to the TNF-α receptors and several lines of evidence indicate that it may transduce a cell death signal in various types of nerve cells (Rabizadeh et al. 1993; Barrett and Bartlett 1994; Von Bartheld et al. 1994; Frada et al. 1996; Van der Zee et al. 1996). Expression of the LNGFR dramatically declines during fetal or postnatal life in numerous CNS structures, such as the spinal cord, and 1,25–D3 amplifies this decrease (Naveilhan et al, 1996 a). This effect resembles that elicited by a T3 treatment in the cerebellum (Figueiredo et al. 1993). It is possible, therefore, that through its regulatory action on the LNGFR receptor, 1,25-D3 might influence the susceptibility of neuronal populations to cell death signals.

1,25-D3 Controls the Supply in Neurotrophic Factors

Our in vitro studies, performed on newborn rat primary astrocytes, showed that the treatment of these cells with the hormone induces the synthesis of its own receptor, the VDR, and that of a vitamin D catabolizing enzyme, the 1,25-D3-24-hydroxylase (Naveilhan et al. 1993). An important response of astrocytes to 1,25–D3 is an enhanced synthesis of NGF (Wion et al. 1991; Neveu et al. 1994b). This promoting effect also takes place in vivo, in the cortex or hippocampus of rats undergoing a systemic treatment with 1,25-D3 (Saporito et al. 1994). Enhancement of NGF synthesis accounts for the augmentation of choline-acetyltransferase specific activity in the brain of 1,25-D3-treated rats (Sonnenberg et al. 1986). However, 1,25-D3 also induces in astrocytes an accumulation of transcripts encoding neurotrophin 3 (NT3) or glial cell line-derived neurotrophic factor (GDNF) in C6 glioma cells (Neveu et al. 1994c; Naveilhan et al. 1996b). Therefore, 1,25-D3 can exert a neuroprotective effect through its ability to enhance the production of several neurotrophic factors.

Immunosuppressive Actions of 1,25-D3 in the Rat CNS During Inflammatory Reactions

In order to explore the mechanisms by which 1,25-D3 can exert immunosuppressive effects in the CNS, two rat models of brain inflammation were used in our laboratory: relapsing experimental autoimmune encephalomyelitis (EAE) and a purely inflammatory reaction consisting of the injection of lipopolysaccharide (LPS) into the rat hippocampus.

Experimental Autoimmune Encephalomyelitis

EAE was induced in Lewis rats by immunization with a guinea pig spinal cord extract. Under such experimental conditions, animals develop a cell-mediated autoimmune disease that affects the CNS. The clinical course of this chronic relapsing EAE is characterized by two paralytic attacks, which culminate 12 and 23 days post-immunization, separated by a remission phase that occurs around day 18 post-immunization. In this system, infiltrating cells comprise monocytes/macrophages and T lymphocytes. Inflammatory infiltrates are first seen in the spinal cord, but extend to other structures such as the cerebellum during the relapse. To investigate the therapeutic potentialities of 1,25-D3 on an *ongoing* autoimmune disease, animals were treated by intraperitoneal route *after the beginning of clinical signs* (5 μg/kg on days 11 and 13 post-immunization, then 1 μg/kg on days 19, 21 and 23). This treatment resulted in a significant improvement of the pathology (Nataf et al. 1996), a fact supported by a simultaneous study performed in a mouse model (Cantorna et al. 1996). The action of 1,25-D3 appeared to differ depending on the brain areas. Immunohistochemical analysis showed that infiltrates of OX42-positive cells (monocytes/macrophages) appeared less abundant in the anterior brain of 1,25-D3-treated EAE rats. In these OX42-positive cells, the expression of CD4 antigen was significantly reduced in all structures examined, with an especially pronounced effect in the brainstem and cerebellum. A significant decrease in the expression of MHC class II molecules was also observed in anterior brain and midbrain sections (Nataf et al. 1996). This finding might explain, at least in part, the improvement of clinical signs elicited by 1,25–D3, since the decreased expression of the CD4 and MHC class II molecules should limit the ability of monocytes, macrophages and microglial cells to function as antigen-presenting cells.

Type II NO-synthase (NOS II), also referred to as inducible NO-synthase (iNOS), catalyzes the generation of a diffusible, short-lived molecule, nitric oxide (NO). The enzyme is expressed by activated monocytes/macrophages or microglial cells, while a closely related isoform is expressed by reactive astrocytes (Galea et al. 1994). NOS II is induced in the CNS in response to various types of lesions or disorders, such as ischemia, Alzheimer's and Parkinson's disease, but also multiple sclerosis and EAE (Bagasra et al. 1995; Bö et al. 1994; Endoh et al. 1994; Garcion et al. 1997; Hunot et al. l996; Vodovots et al. 1996). In vitro approaches indicate that NOS II can be induced by a large variety of pro-inflammatory stimuli, such as TNF-α, IL-1β or IFN-γ. Conversely, NOS II expression is reduced in response to immunosuppressive lymphokines such as IL-4, IL-10 or TGF-β (Szabo and Thiemermann 1995). NO exerts numerous physiological functions but, when produced in excess, it may become toxic to neurons or oligodendrocytes (Merrill et al. 1993; Dawson and Dawson 1996). It is currently believed that NOS II is closely associated with demyelinating phenomena occurring during multiple sclerosis or with neuronal death following ischemia (Bö et al. 1994; Endoh et al. 1994).

On this basis, it was interesting to investigate whether 1,25-D3 could modify the expression of NOS II during an ongoing EAE. In the absence of treatment, double-labeling immunohistochemical approaches and in situ hybridization experiments demonstrated that the cells of the monocyte/macrophage lineage continuously contributed to NOS II expression (from day 12 to day 23 post-immunization), whereas astrocytes only expressed NOS II immunoreactivity and mRNA during the relapse (day 23). Treatments with 1,25-D3, carried out as described above, resulted in a drastic inhibition of NOS II expression, both at the mRNA and protein levels. This effect was also region-specific and concerned mainly the cerebellum and brainstem. NOS II down-regulation, which occurred both in cells of the monocyte lineage and in astrocytes, could also account for the beneficial effects of a curative treatment with 1,25-D3 (Garcion et al. 1997).

Focal Inflammatory Reaction Induced by LPS Intracerebral Injection

NOS II regulation was also studied in a model of brain inflammation obtained following an intrahippocampal injection of LPS, which triggers a dramatic mobilization of cells belonging to the monocytic lineage, without any lymphocyte recruitment (Montero-Menei et al. 1994, 1996). LPS intracerebral injection also induced NOS II mRNA and protein. This early, transient and restricted expression occurred first in choroid plexus and ependymal cells (5 hours post injection), then in macrophages and activated/reactive microglia (15–24 hours), and finally in astrocytes (from 72 hours onwards). To investigate potential regulatory effects of 1,25-D3 on NOS II expression, we delivered this hormone concomitantly with LPS into the rat hippocampus (Garcion et al. 1998). As observed in EAE (see above), this treatment significantly inhibited NOS II expression in monocytic/microglial cells at the mRNA (Fig. 1) and protein levels. In contrast, no effect on the astrocytic enzyme could be evidenced. Since NOS II immunoreactivity was only detectable in these cells from three days onwards, it is likely that 1,25-D3, injected at the beginning of the experiment, was already degraded after such a delay. However, 72 hours after the injection of LPS and 1,25-D3 we noticed a sixfold increase in the number of macrophages around the lesion site, which correlated with a decrease in the proportion of apoptotic cells.

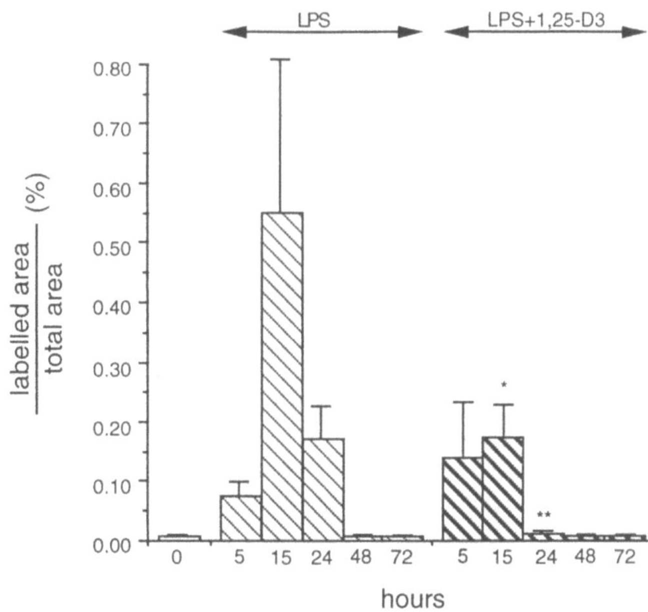

Fig. 1. Expression of NOS II mRNA shown by in situ hybridization analysis in control rat brain (0) and 5, 15, 24, 48 and 72 hours after injection of LPS, alone or in combination with 1,25-D3. The labelled areas, corresponding to NOS II mRNAs, were quantified by densitometric scanning (Vision Lab 1000 system, Biocom, Les Ullis, France). Student's *t* test: *P <0.05 ; ** P <0.01

These data demonstrate, in two in vivo experimental models, that 1,25-D3 is a potent regulator of NOS II expression. Therefore, we have investigated in vitro whether the hormone could directly interfere with the expression of the NOS II gene. While studies on microglial cells are currently in progress, data obtained in astrocyte cultures stimulated with LPS did not evidence any down-regulation of NOS II transcripts by 1,25–D3 (Garcion et al. unpublished data). In contrast, the hormone decreased the pool of transcripts encoding M-CSF (CSF-1) or TNF-α (Furman et al. 1996). This lack of direct action on NOS II transcripts suggests that, in vivo, 1,25-D3 acts as a regulator of inflammatory reactions through a complex network, by modifying the balance between anti-inflammatory and pro-inflammatory cytokines.

Conclusion

1,25-D3 appears to act as a multifunctional mediator controlling the homeostasis of the CNS. The hormone has the potential to increase the production of NGF and possibly that of other neurotrophic factors. Likewise, 1,25-D3 elicits anti-inflammatory effects in astrocytes, microglial cells and infiltrating monocytes/macrophages. It is interesting to note, in this respect, that vitamin D deprivation increases the clinical impairments caused by an ongoing EAE in mice (Cantorna et al. 1996) and alters glucose transport and metabolism in the cortex (Stio et al. 1993). In human, vitamin D deficiencies are frequent in home-bound elderly persons, as well as in multiple sclerosis patients (Gloth et al. 1995; Nieves et al. 1994) and they constitute a risk factor for pathological complications in AIDS patients (Haug et al. 1995). In Alzheimer's disease, there seems to be a selective reduction of VDR mRNA or calbindin D28k protein in hippocampal neurons (Sutherland et al. 1992; Maguire-Zeiss et al. 1995). These data underscore the interest in adequate supplementation of vitamin D. It should be stressed, however, that 1,25-D3 controls calcemia and that supraphysiologic concentrations are highly toxic. Furthermore, neuromodulatory or immunomodulatory effects of 1,25-D3 are often, but not exclusively, observed at so-called pharmacological doses which are above the physiological ones. Several 1,25-D3 analogues have been synthesized which display a much lower hypercalcemic effect than that elicited by the natural hormone, but which keep the potential to exert nonconventional effects (Lemire et al. 1994; Norman 1995). These compounds, if they keep the regulatory properties of 1,25-D3 in neuroimmune phenomena, may turn out to be of valuable help in the therapy of neurodegenerative diseases.

Acknowledgments. This work was supported by the Institut National de la Santé et de la Recherche Médicale, the Ligue Française contre la Sclérose en Plaques, and the Association de Recherche contre la Sclérose en Plaques (grant to E.G.).

References

Alexianu MB, Robbins B, Carswell S, Appel SH (1998) 1α, 25-Dihydroxyvitamin D3-dependent up-regulation of calcium-binding proteins in motoneuron cells. J Neurosci Res 51:58–66

Bagasra O, Michaels FH, Zheng YM, Bobroski LE, Spitsin SV, Fang Fu Z, Tawadros R, Koprowski H (1995) Activation of the inducible form of nitric oxide synthase in the brains of patients with multiple sclerosis. Proc Natl Acad Sci USA 92:12041–12045

Balabanova S, Richter H-P, Antoniadis G, Homoki J, Kremmer N, Hanle J, Teller WM (1984) 25-Hydroxyvitamin D, 24,25-dihydroxyvitamin D and 1,25-dihydroxyvitamin D in human cerebrospinal fluid. Klin Wochenschr 62:1086–1090

Barrett GL, Bartlett PF (1994) The p75 nerve growth factor receptor mediates survival or death depending on the stage of sensory neuron development. Proc Natl Acad Sci USA 91:6501–6505

Bidmon HJ, Mayerhofer A, Heiss C, Bartke A, Stumpf WE (1991) Vitamin D (soltriol) receptors in the choroid plexus and ependyma: their species-specific presence. Mol Cell Neurosci 2:145–156

Bö L, Dawson TM, Wesselingh S, Mork S, Choi S, Kong PA, Hanley D, Trapp BD (1994) Induction of nitric oxide synthase in demyelinating regions of multiple sclerosis brains. Ann Neurol 36:778–786

Cantorna MT, Hayes CE, DeLuca HF (1996)1,25-Dihydroxyvitamin D3 reversibly blocks the progression of relapsing encephalomyelitis, a model of multiple sclerosis. Proc Natl Acad Sci USA 93:7861–7864

Christakos S, Raval-Pandya M, Wernyj RP, Yang W (1996) Genomic mechanisms involved in the pleiotropic actions of 1,25-dihydroxyvitamin D3. Biochem J 316:361–371

Clavreul A, D'Hellencourt CL, Montero-Menei C, Potron G, Couez D (1998) Vitamin D regulates B7.1 and B7.2 expression on human peripheral blood monocytes. Immunology 95:272–277

Clemens TL, Garrett KP, Zhou XY, Pike WJ, Haussler MR, Dempster DW (1988) Immunocytochemical localization of the 1,25-dihydroxyvitamin D3 receptor in target cells. Endocrinology 122:1224–1230

Dawson VL, Dawson TM (1996) Nitric oxide neurotoxicity. J Chem Neuroanat 10:179–190

Endoh M, Maiese K, Wagner J (1994) Expression of the inducible form of nitric oxide synthase by reactive astrocytes after transient global ischemia. Brain Res 651:92–100

Figueiredo BC, Almazan G, Ma Y, Tetzlaff W, Miller FD, Cuello AC (1993) Gene expression in the developing cerebellum during perinatal hypo- and hyperthyroidism. Mol Brain Res 17:258–268

Frada JM, Rodriguez-Tebar A, Barde YA (1996) Induction of cell death by endogenous nerve growth factor through its p75 receptor. Nature 383:166–168

Furman I, Baudet C, Brachet P (1996) Differential expression of M-CSF, LIF, and TNF-α genes in normal and malignant glial cells: regulation by lipopolysaccharide and vitamin D. J Neurosci Res 46:360–366

Galea E, Reis DJ, Feinstein DL (1994) Cloning and expression of nitric oxide synthase from rat astrocytes. J Neurosci Res 37:406–414

Garcion E, Nataf S, Bérod A, Darcy F, Brachet P (1997) 1,25-Dihydroxyvitamin D3 inhibits the expression of inducible nitric oxide synthase in rat central nervous system during experimental allergic encephalomyelitis. Mol Brain Res 44:255–267

Garcion E, Do Thanh X, Bled F, Teissier E, Dehouck MP, Rigault F, Brachet P, Girault A, Torpier G, Darcy F (1996) 1,25-Dihydroxyvitamin D3 regulates γ-glutamyl transpeptidase activity in rat brain. Neurosci Lett 216:183–186

Garcion E, Sindji L, Montero-Menei C, André C, Brachet P, Darcy F (1998) Expression of inducible nitric oxide synthase during brain inflammation; regulation by 1,25-dihydroxyvitamin D3. Glia 22:282–294

Gascon-Barré M, Huet PM (1983) Apparent (3H)1,25-dihydroxyvitamin D3 uptake by canine and rodent brain. Am J Physiol 44:E266–E271

Gloth III FM, Gundberg CM, Hollis BW, Haddad JG, Tobin JD (1995) Vitamin D deficiency in homebound elderly persons. JAMA 274:1683–1686

Haug CJ, Aukrust P, Lien E, Müller F, Espevik, T, Froland S (1995) Disseminated Mycobacterium avium complex infection in AIDS: immunopathogenic significance of an activated tumor

necrosis factor system and depressed serum levels of 1,25-dihydroxyvitamin D. J Infect Dis 173:259–262
Haussler MR, Whitfield GK, Haussler CA, Hsieh JC, Thompson PD, Selznick SF, Encinas Dominguez C, Jurutka PW (1998) The nuclear vitamin D receptor: biological and molecular regulatory properties revealed. J Bone Min Res 13:325–349
Hunot S, Boissière F, Faucheux B, Brugg B, Mouatt-Prigent A, Agid Y, Hirsch EC (1996) Nitric oxide synthase and neuronal vulnerability in Parkinson's disease. Neuroscience 72:355–363
Kreutz M, Andreesen R (1990) Induction of human monocyte to macrophage maturation in vitro by 1, 25-dihydroxyvitamin D3. Blood 76:2457–2461
Lemire JM, Archer CD (1990) 1,25-dihydroxyvitamin D3 prevents the in vivo induction of murine encephalomyelitis. J Clin Invest 87:1103–1107
Lemire JM, Ince A, Takashima M (1992) 1,25-Dihydroxyvitamin D3 attenuates the expression of experimental murine lupus of MRL/l mice. Autoimmunity 12:143–148
Lemire JM, Archer DC, Beck L, Reddy GS, Uskokovic MR, Spiegelberg HL (1994) The role and mechanism of vitamin D analogs in immunosuppression. In: Norman AW, Bouillon R, Thomasset M (eds) Vitamin D, a pluripotent steroid hormone: structural studies, molecular endocrinology and clinical applications. Walter de Gruyter, Berlin, New York, pp 531–539
Maguire-Zeiss KA, Li ZW, Shimoda LMN, Hamill RW (1995) Calbindin D28k mRNA in hippocampus, superior temporal gyrus and cerebellum: comparison between control and Alzheimer disease subjects. Mol Brain Res 30:362–366
Mathieu C, Laureys J, Sobis H, Vandeputte M, Waer M, Bouillon R (1992) 1,25-Dihydroxyvitamin D3 prevents insulitis in NOD mice. Diabetes 41:1491–1495
Merrill JE, Ignarro LJ, Sherman MP, Melinek J, Lane TE (1993) Microglial cell cytotoxicity of oligodendrocytes is mediated through nitric oxide. J Immunol 151:2132–2141
Montero-Menei C, Sindji L, Pouplard-Barthelaix A, Jehan F, Dénéchaud L, Darcy F (1994) Lipopolysaccharide intracerebral administration induces minimal inflammatory reaction in rat brain. Brain Res 653:101–111
Montero-Menei C, Sindji, Garcion E, Mège M, Couez D, Gamelin E, Darcy F (1996) Early events of the inflammatory reaction induced in rat brain by lipopolysaccharide intracerebral injection relative contribution of peripheral monocytes and activated microglia. Brain Res 724:55–66
Nataf S, Garcion E, Darcy F, Brachet P (1996) 1,25-Dihydroxyvitamin D3 exerts regional effects in the central nervous system during experimental allergic encephalomyelitis. J Neuropathol Exp Neurol 55:904–914
Naveilhan P, Neveu I, Baudet C, Ohyama KY, Brachet P, Wion D (1993) Expression of 25 (OH) vitamin D3 24-hydroxylase gene in glial cells. NeuroReport 5:255–257
Naveilhan P, Neveu 1, Baudet C, Funakoshi H, Wion D, Brachet P, Metsis M (1996a) 1,25-Dihydroxyvitamin D3 regulates the expression of the low-affinity neurotrophin receptor. Mol Brain Res 41:259–268
Naveilhan P, Neveu I, Wion D, Brachet P (1996b) 1,25-Dihydroxyvitamin D3, an inducer of glial cell line-derived neurotrophic factor. Neuroreport 7:2171–2175
Neveu I, Naveilhan P, Menaa C, Wion D, Brachet, P, Garabedian M (1994a) Synthesis of 1,25-dihydroxyvitamin D3 by rat brain macrophages in vitro. J Neurosci Res 38:214–220
Neveu I, Naveilhan P, Jehan F, Baudet C, Wion D, De Luca HF, Brachet P (1994b) 1,25-Dihydroxyvitamin D3 regulates the synthesis of nerve growth factor in primary cultures of glial cells. Mol Brain Res 24:70–76
Neveu I, Naveilhan P, Baudet C, Brachet P, Metsis M (1994c) 1,25-dihydroxyvitamin D3 regulates NT-3, NT-4 but not BDNF mRNA in astrocytes. NeuroReport 6:124–126
Nieves J, Cosman F, Herbert J, Shen V, Lindsay R (1994) High prevalence of vitamin D deficiency and reduced bone mass in multiple sclerosis. Neurology 44:1687–1692
Norman AW (1995) The vitamin D endocrine system: Manipulation of structure-function relationships to provide opportunities for development of new cancer chemopreventive and immunosuppressive agents. J Cell Biochem 22:218–225
Pardridge WM, Sakyiama R, Loty WA (1985). Restricted transport of vitamin D and A through the rat blood-brain barrier. J Neurochem 44:1138–1140

Provvedini DM, Deftos LJ, Manolagas SC (1986) 1,25-Dihydroxyvitamin D3 promotes in vitro morphologic and enzymatic changes in normal human monocytes consistent with their differentiation into macrophages. Bone 7:23–28

Rabizadeh S, Oh J, Zhong L, Yand J, Bitler CM, Butcher LL, Bredesen DE (1993) Induction of apoptosis by the low-affinity NGF receptor. Science 261:345–348

Reichel H, Koeffler HP, Tobler A, Norman AW (1987) 1α,25-Dihydroxyvitamin D3 inhibits γ-interferon synthesis by normal human peripheral blood lymphocytes. Proc Natl Acad Sci USA 84: 3385–3389

Rigby WF, Denome S, Fanger MW (1987) Regulation of lymphokine production and human T lymphocyte activation by 1,25-dihydroxyvitamin D3. J Clin Invest 79:1659–1664

Rigby WF, Waugh M, Graziano RF (1990) Regulation of human monocyte HLA-DR and CD4 antigen expression by 1,25-dihydroxyvitamin D3. Blood 76:189–197

Rook GAW, Steele J, Faher L, Barker S, Karmali R, O'Riordan J, Stanford J (1986) Vitamin D3, gamma interferon, and control of proliferation of mycobacterium tuberculosis by human monocytes. Immunology 57:159–163

Saporito MS, Brown ER, Hartpence KC, Wilcox HM, Vaught JL, Carswell S (1994) Chronic 1,25-dihydroxyvitamin D3-mediated induction of nerve growth factor mRNA and protein in L929 fibroblasts and in adult rat brain. Brain Res 633:189–196

Sonnenberg J, Luine VN, Krey LC, Christakos S (1986) 1,25-Dihydroxyvitamin D3 treatment results in increased choline acetyltransferase activity in specific brain nuclei. Endocrinology 118:1433–1439

Spittler A, Willheim M, Leutzmezer F, Ohler R, Krugluger W, Reissner C, Brodowicz T, Roth T, Boltz-Nitulescu G (1997) Effect of 1α, 25 dihydroxyvitamin D3 on the expression of MHC antigens, complement receptors and other antigens on human blood monocytes and U937 cells: role in cell differentiation, activation and phagocytosis. Immunology 90:286–293

Stio M, Lunghi B, Vincenzini MT, Treves C (1993) Effect of vitamin D deficiency and 1,25-dihydroxyvitamin D3 on metabolism and D glucose transport in rat cerebral cortex. J Neurosci Res 35:559–566

Stumpf WE, O'Brien LP (1987) 1,25-(OH)2 Vitamin D3 sites of action in the brain. An autoradiography study. Histochemistry 87:393–406

Sutherland MK, Somerville MJ, Yoong LKK, Bergeron C, Haussler MR, Crapper Mc Lachlan DR (1992) Reduction of vitamin D hormone receptor mRNA levels in Alzheimer as compared to Huntington hippocampus: correlation with calbindin-28k mRNA levels. Mol Brain Res 19:239–250

Szabo C, Thiemermann C (1995) Regulation of the expression of the inducible isoform of nitric oxide synthase. Adv Pharmacol 34:113–153

Van der Zee CEEM, Ross GM, Riopelle RJ, Hagg T (1996) Survival of cholinergic neurons in developing p75 NGFR-deficient mice. Science 274:1729–1732

Vodovots Y, Lucia MS, Flanders KC, Chesler L, Xie Q-W, Smith TW, Mumford R, Webber R, Nathan C, Roberts AB, Lippa CF, Sporn MB (1996) Inducible nitric oxide synthase in tangle-bearing neurons of patients with Alzheimer's disease. J Exp Med 184:1425–1433

Von Bartheld CS, Kinoshita Y, Prevette D, Yin QW, Oppenheim RW, Bothwell M (1994) Positive and negative effects of neurotrophins in the isthmo-optic nucleus in chick embryo. Neuron 12:639–654

Wion D, MacGrogan D, Neveu I, Houlgatte R, Brachet P (1991) 1,25-Dihydroxyvitamin D3 is a potent inducer of nerve growth factor synthesis. J Neurosci Res 28:110–114

Zarrabeitia MT, Riancho JA, Amado JA, Olmos JM, Gonzalez-Macias J (1992) Effect of calcitriol on the secretion of prostaglandin E2, interleukin 1, and tumor necrosis factor α by human monocytes. Bone 13:185–189

Subject Index

MIX
Papier aus verantwortungsvollen Quellen
Paper from responsible sources
FSC® C105338

FSC
www.fsc.org

If you have any concerns about our products,
you can contact us on
ProductSafety@springernature.com

In case Publisher is established outside the EU,
the EU authorized representative is:
Springer Nature Customer Service Center GmbH
Europaplatz 3, 69115 Heidelberg, Germany

Printed by Libri Plureos GmbH
in Hamburg, Germany